IWER THOR LORENZEN (1
in Harrislee Flensburg, Germ
career as a teacher in 1914. Whilst serving in the First World War, he became acquainted with anthroposophy—as founded by Rudolf Steiner—through a fellow soldier. After the war he returned to teaching, later moving into special education. He set up his own school near Hamburg in 1949, where he remained until his retirement. Having worked as a volunteer in the Zoological State Institute in Hamburg from 1935 onwards, Lorenzen was also a biologist who was revered for his love and knowledge of beekeeping, particularly amongst biodynamic farmers. He published his key work on beekeeping in 1938 and wrote another nine books as well as numerous articles on the insect and animal world, metamorphosis and evolution.

The Spiritual Foundations of Beekeeping

Iwer Thor Lorenzen

Translated from German by Paul King

Temple Lodge Publishing Ltd.
Hillside House, The Square
Forest Row, RH18 5ES

www.templelodge.com

Published in English in 2017 by Temple Lodge Publishing in association with the Natural Beekeeping Trust www.naturalbeekeepingtrust.org

Originally published in German under the title *Die geistigen Grundlagen der Bienenzucht* by Schultz & Thiele, Hamburg 1938

A CIP catalogue record for this book is available from the British Library

ISBN 978 1 906999 98 8

Cover by Morgan Creative featuring a painting © Arif Turan
Typeset by DP Photosetting, Neath, West Glamorgan
Printed and bound by 4Edge Ltd., Essex

Contents

Foreword	1
Preface	12
Introduction	15
Relation of the honeybee to the floral kingdom and the developmental level of digestion in the bee	17
On the Treatment of Bee Diseases	27
The Origin of the Honeybee	35
The Bee Colony as Individuality and Group-soul	50
The Question of Appropriate Beekeeping Techniques	74
Afterword	84
Notes	87

Foreword

The Natural Beekeeping Trust is honoured to present an English translation of Iwer Thor Lorenzen's *The Spiritual Foundations of Bee Husbandry*. First published some 80 years ago, this book will speak to all who perceive the physical world as one aspect of a much larger totality of existence, in which everything has its being, and which is accessible to those who are prepared to make the effort. Lorenzen makes the effort and shares the fruits of his labours, as does the translator, Paul King, to whom credit and thanks are due.

This book is about a radically different approach to bees and beekeeping and the need for that is more urgent now than it has ever been. At the time of Lorenzen's writing, only a few far-sighted observers could see the harm that would be caused by many of the practices that were then taking hold in beekeeping. Now it is painfully apparent across the globe that something is very wrong, not just with our approach to bees but with our whole approach to agriculture and the ecosystem in which we live and on which we depend. The resultant crisis highlights a void in our relationships and understanding. If the translation of this little book contributes, even in small measure, to filling that void, and creating a fuller understanding of the world and the place within it of the magnificent honeybee, it will have served its purpose.

Lorenzen builds on the work of Rudolf Steiner. Steiner in turn draws on certain concepts originally developed by J. W. von Goethe. Seeing the world through Goethean eyes is very different from seeing the world in reductionist, mechanistic terms. Beekeeping based on this latter approach perceives the honeybee colony as a set of physical components that can be taken apart and swapped around at will by the beekeeper, even to the point of being moved from one colony to another. This fails completely to see that the honeybee cannot be understood by physical dissection, whether that be of the colony or of the individual bee. To understand the bee one has to step back and see the whole, not just as an integrated being, but as an integrated being set in its wider context, the world through which it has its existence.

It is this wider viewpoint that is so sadly lacking in the approach of the greater part of current beekeeping. Lorenzen makes clear in his introduction that he was keenly aware of this even 80 years ago. For him, wholeness, context and relationship are key to understanding. When he addresses the evolution of the bee, he does so in the context of its nutritional and physiological relationship with the flower and he is prescient in understanding aspects of nutrition that have only very recently come to the attention of conventional science. Lorenzen also covers the relationship of the bee with some of its diseases, the relationship of the bee with the beekeeper and, of key importance, the relationship of the bee with itself. In the latter case, the question Lorenzen addresses is how, given its different levels of existence—individual bees, bee colonies and the honeybee species—the bee has a sense of self. Not surprisingly perhaps, we find that the queen is pivotal, and we learn why it is that certain beekeeper actions can be harmful when performed at certain times of the colony cycle but are less so at others.

The reductionist world view is often epitomized by reference to the English physicist Isaac Newton. Although Newton himself studied in depth the work of the alchemists, he is most remembered for his work as an astronomer and physicist. As part of his study of light, Newton conducted a now famous experiment in which he passed a beam of light through a prism and thereby produced a rainbow of colours. The theory that Newton developed to explain this phenomenon is today well known and is taught in all physics classes. It is that white light—or more correctly, colourless light—is not really white at all but contains within it all the colours of the rainbow. A prism can cause these colours to separate, making them visible.

Newton published his thesis on light in 1704. Just over 100 years later, Goethe, the German playwright and polymath, published his own *Theory of Colours*. Goethe comes to very different conclusions about colour from those of Newton. The origins of this difference can be traced to the different ways in which the two men approached the subject. While Newton focussed entirely on light, Goethe looked at the matter in the round. Moreover, while Newton sought to eliminate the observer from the frame, Goethe did no such thing. He accepted the observer as part and parcel of the set up. Thus, Newton carefully arranged for light to shine as a narrow beam through a prism in a darkened room before impinging on a screen. Goethe simply picked up a prism and looked through it at a brightly lit window. In Newton's arrangement, one sees colours on the screen. In Goethe's case one sees no colours, just the window and what lies beyond. Only when one looks at the silhouette of one of the glazing bars or the dark edge of the window do colours appear. As Goethe put it:

> Along with the rest of the world I was convinced that all the colours are contained in the light; no one had ever told me anything different, and I had never found the least cause to doubt it, because I had no further interest in the subject.
>
> But how I was astonished, as I looked at a white wall through the prism, that it stayed white! That only where it came upon some darkened area, it showed some colour, then at last, around the window sill all the colours shone.[1]

In other words, the world you see depends entirely on where you choose to stand and how you choose to view. Standing outside and interrogating the world as 'other', as Newton did, will give a quite different outcome from being part of the world and interacting with it as Goethe did. Newton saw light as many beekeepers see their bee colonies: something to be interrogated by dissection. He concentrated solely on the beam of light, and saw all the other elements in the room as tools for interrogating and dissecting that beam. He concluded that colour was contained in the beam of light. Where else could it be if all else is ignored?

By contrast, Goethe looked at all of the elements in the room with which light interacted and considered them in relation one to another. As a result, rather than seeing colour as intrinsic to light, he concluded that it was something created *ab initio* by the interaction of all the various elements in the room, light, dark and his own eye. In other words, the two men came to diametrically opposite conclusions due to the different contexts in which they observed the phenomenon. One saw colour as a pre-existing, static, component of light, visible on dissection. The other saw colour as a dynamic, context-dependent, creation.

Goethe is sometimes criticized for relying too heavily on empiricism without sufficiently developing a theory for the underlying mechanism. This, of course, assumes two things. The first is that there is a 'mechanism' that underlies the phenomenon. The second is that any such mechanism is capable of being reduced to terms that can be written down, often in the form of mathematics. Goethe preferred to patiently observe, as a naturalist or painter would observe, without preconceptions. He believed in looking and seeing as a form of direct experience, something he described as 'a delicate empiricism which makes itself utterly identical with the object ...'.[2] This applies in particular to the observation of nature:

> Natural objects should be sought and investigated as they are and not to suit observers, but respectfully as if they were divine beings.[3]

Moreover, Goethe preferred not to look beyond the phenomenon to construct intellectual edifices that served to explain the observations. The danger in so doing is that the intellectual edifices come to supplant the phenomenon under investigation, such that the unobserved replaces the observed. In the words of Henri Bortoft:

> Goethe's way of thinking is concrete, not abstract, and can be described as one of dwelling in the phenomenon.[4]

The point of dwelling in the phenomenon is that, through the process of 'exact sensorial imagination'[5]—taking the phenomenon into one's mind in a sensory way and allowing it to dwell there—one comes to know and understand the phenomenon in its own terms. This does, of course, require a setting aside of the observer's ego. The referential framework within which the observations take place must come from the phenomenon itself, rather than from the observer. This is not to say that all externalities must be ignored; quite the opposite. They are fully taken into account but this is done from the perspective of the phenomenon rather than from the imposed perspective of the observer. For the observer to avoid inadvertently importing bias into his or her observations, it is vital that all pre-conceptions be laid aside. Before this can happen such preconceptions must, of course, be recognized. This in itself can be tricky, as many preconceptions, one might even say most, are so deeply held, so ingrained, as to be unconscious. Setting aside preconceptions is far from easy and a certain amount of time, and many attempts, may pass before the observer begins to get the feel—and feel is probably the most apposite term—for, and of, the phenomenon in its own right. In the words of Charles Martin Simon: 'Forget everything you ever learned and start observing what is really going on'.[6]

Apart from those areas of quantum physics where scientists wonder whether certain quantum phenomena exist in the absence of an observer, the current approach of reductionist science to the observer is to regard him or her as a nuisance. It is posited that observers introduce bias and, hence, only those facts that can be measured by instruments carry sway. This approach leads ultimately to a decontextualization of the phenomenon being studied. Any conclusions drawn from the observations are also decontextualized and can, as a result, be very misleading.

The Goethean view is that the observer is a reality and has a role both as observer and as participator; he or she is part of the creative dynamic. In contrast, rather than accepting the subtle and deep interaction that can exist between the bees and the beekeeper, the approach of most beekeeping over many decades has been to pretend that the beekeeper is somehow external, with the bees being viewed as a mechanical, dissectible, construct. This viewpoint was already well established at the time Lorenzen was writing. Some twelve years before, in his lectures on bees at Dornach, Rudolf Steiner cautioned against such a reductionist paradigm:

> With bees, this is the very thing you must pay attention to: that the object of our concern has nothing to do with the individual bees but rather with that which absolutely belongs together to make up the whole. This is something that a simple understanding of the matter can't fathom. In this case you must be able to view the whole as it really must be. That is why we can learn so much from a beehive or things like it, because a beehive completely contradicts the thoughts we may form about it.[7]

In other words, one must see the whole, and the relationship of the whole to the outside world, in its own terms. One must not come with preconceptions and one must be prepared to have any such preconceptions challenged and contradicted. In short, one must be prepared to see a quite different way of being from that to which one is accustomed. To paraphrase Steiner, this is not a matter that a simplistic understanding can fathom. We need instead to prepare ourselves to go to a place to which we have not previously ventured and, when we get there, we need to accept that matters will be complex and often counter-intuitive. A simple reductionist view will inevitably fail in this endeavour, not least because it is simple and reductionist.

To appreciate the bee one must be prepared to accommodate complexity and subtlety and to enter a world where the forces and dynamics are very different from those that apply to the normal spheres of human activity. One needs to leave behind one's normal view of the world and enter another view, gradually and respectfully. One must be mindful that the bee is a highly evolved creature that embodies within it aeons of wisdom that has been winnowed and refined by subtle forces and influences beyond our everyday experience. It may sound from this that one can never experience the world of the bee. In a sense, this is true. But, in another sense, we can try. We can put ego to one side and develop a faculty for Goethe's exact sensorial imagination. We can then start to go beyond ourselves, to enter a different realm.

This is precisely what Lorenzen has done. He takes us through various aspects of the world seen, as far as is possible, from the perspective of the bee. Given the importance of context and relationships, it is not surprising that Lorenzen starts by examining the relationship between bees and plants.

A discussion of growth forces in plants and insects culminates in a consideration of why it is that bees feed exclusively on flowers. The flower is the end stage of the plant life cycle and pollen and nectar are the end products of the flower. As such, they have less of the essential nature of the plant about them than is present in other parts, such as the leaf. Lorenzen describes the need for bees to feed exclusively on such end products as indicating a weakening of the bees' digestive forces when it comes to coping with the essence of the plant. Pollen and nectar require little transformation before they can be assimilated into the bee body. In Lorenzen's words:

> It lies within the healthy forces of the honeybee to start its digestive process at the point where the plant leaves off in the flower. The metabolic organisation of the honeybee is incomplete without the processes of the flower. Only together do they form a totality.

Recent research takes this analysis even further. Pollen and nectar nearly always contain a phytochemical with the name coumaric acid. When a queen bee lays a fertilized egg in a queen cell, the standard explanation is that a surfeit of particular nutrients contained in royal jelly is all that is required for that egg to develop into a queen. However, it now seems that what is excluded from the diet is also important. If coumaric acid is present, the queen's ovaries do not fully develop.[8] Pollen and nectar containing coumaric acid are added by nurse bees to the diet of larvae destined to be worker bees but are assiduously excluded from the diet of larvae destined to be queen bees. So, if the quintessential component of the bee, the queen, is to be fully developed, she needs to be shielded even from certain plant materials that are in the flower.

The discussion of bee digestion leads in the following chapter to a consideration of certain ailments of bees, in particular dysentery and nosema. Lorenzen points us to herbal remedies and mentions planting regimes. He also covers various aspects of bee physiology. Not surprisingly, he deals with the latter in a holistic manner, showing how the various parts work together in harmony to create health through dynamic rhythm and balance, which he advises the beekeeper to be wary of upsetting through injudicious feeding.

The next chapter deals with the origin of the honeybee. Lorenzen makes the contextual point that how one sees bee evolution will depend very much on one's perspective. Just as we saw in the case of colour, a conclusion based on a close-up view may be very different from one that is obtained by standing back and looking at matters on a much larger scale and in a much larger context. This is especially so when one includes in one's perspective behaviours and other matters which exhibit themselves in ways that cannot be seen if one restricts oneself solely to questions of physical form.

Lorenzen first takes us through the Darwinian view of honeybee evolution that, based on shared physical characteristics, bees are thought to have evolved out of a group of wasps broadly known as digger wasps. He then introduces us to the fig-wasp and deals with the development of both that wasp and cultivated figs in great detail, illustrating the identical functionality as between the fig-wasp in its role of pollinating figs and the honeybee in its role of more general pollination. On this basis, Lorenzen supports the view put forward by Steiner that the honeybee descends not from the digger wasps but from the fig-wasps.

From here, Lorenzen moves on to a consideration of something that lies at the very heart of the bee, something that, in many ways, is the essence of the bee as an organism. Lorenzen sees the whole colony as surpassing the normal concept of an organism, describing it as 'an organism of the third order'. Today we would say it is a superorganism but, as this term was only coined in 1911 by the American biologist WM Wheeler[9], it is possible that Lorenzen had not come across it.

One can, of course, address the form and function of the superorganism in a reductionist manner by taking concepts that have been applied to lesser organisms and pressing them into service to describe the bee. The result is an aggregation in the mind of the observer of various separate ideas held in unison by a name. What is lacking in this approach is a true understanding of the whole, and that cannot be obtained without a fundamentally different approach to appreciating the bee colony. As Lorenzen himself says of beekeepers and their methods:

> Unfortunately their view of the essential nature of the bee is consistently too superficial for there to be any question of appropriate practices, for it is evident again and again that on essential points only a few have an inkling, let alone a clear consciousness, of the boundaries that are set on beekeeping practices by the nature of the bee.

Although these words were written many years ago, sadly they remain too often true today. Given the importance of context, Lorenzen paints the picture in detail, starting with a lengthy introduction to the evolution of social behaviour in bees generally. He illustrates with many examples the stages between solitary bees and bees, such as bumblebees, that live in colonies. However, for Lorenzen, there is a wholeness about honeybee colonies that stands out as being of a different order, and to this he draws our attention. He examines in turn the queen, the workers, the drones, the colony and the honeybee as a species. He then addresses the all important question: what it is that creates and propagates the essence of the bee and how does it manifest at these various levels?

The answer that Lorenzen gives lies outside the purview of a reductionist mindset. This, however, is not to say that it lies beyond cognizance. On the contrary, anyone who has opened their senses fully to the bee has seen the evidence: the difference in behaviour and mood between a queenright colony and a queenless one; the difference in behaviour and development between a prime swarm and a cast swarm; the difference in behaviour between a swarm and a split; the interaction that exists between a parent colony and its swarms. Apart from the first, these matters will be opaque to beekeepers who suppress that part of the bees' life cycle that has do with swarming and reproduction. They will, however, be plainly apparent to beekeepers who study the whole of the life cycle of the bee from year to year. By carefully observing such essential aspects of the bees' life one can come to see various dynamics at work that encourage further enquiry. That further enquiry Lorenzen carefully lays out, showing us the essential connecting agent that informs and enlivens the processes of the bee as the seasons turn. The key element is shown in physical terms by warmth. Colony warmth levels are indicative of the strength of colony vitality which, in turn, reflects the strength of the colony's vital forces. Lorenzen explains these and other aspects of bee life, not in superficial terms, but in terms of the deeper existence of the bee at the level of the inner worlds. In so doing, he describes a being that exists at levels that are unfathomed by superficial observation. These levels are, nevertheless, within our capacity to understand, and we can connect with them if we open all of our senses in a true Goethean manner.

As has been said, the observer is also a participator and, in the last chapter, Lorenzen turns to the interaction between the bees and the beekeeper. He uses the concepts from the previous chapter to examine commonly practiced beekeeper procedures. Not surprisingly, he finds that many beekeepers fail to understand what it is they are really doing to the bee. This leads to procedures that he says will cause damage in the long term, although they may appear acceptable, or even beneficial, in the short term. Artificial queen rearing by forcible queen removal and the production of queens from worker larvae using the self-healing force of the colony is one such procedure. About this Lorenzen says:

> Anyone who thinks it possible to base and maintain beekeeping on the forces of self-healing *per se*, either has no sense at all for the totality, however constituted, of the hive, or when talking of 'the spirit of the bee' and 'the soul of the colony' is doing so merely in abstract or aesthetic terms.

He examines a series of other procedures and attempts to determine their degree of compliance with his statement in the introduction that 'Beekeeping can only prosper in the long run if it moves in a direction that is determined by the nature of the bee.' About some practices Lorenzen expresses uncertainty and this is wholly in keeping with his aim that this book is a starting place, not an end point.

Gareth John, Natural Beekeeping Trust
December 2016

Preface

The husbandry of plants and animals is something which, apart from the required expert knowledge, also calls for a high degree of attention, conscientiousness and awareness. The more someone is able to exercise loving devotion in their relation to the surrounding world, the more suited they are to being an animal-breeder or plant-grower. Selflessness not only enables them to be more conscientious, but also opens their eyes to the less obvious sides of living nature which must inevitably remain hidden to the dyed-in-the-wool egoist. Furthermore, selfish and selfless individuals do not only differ morally—i.e. in the way they think, feel and will—but even physiologically to the point of the chemical composition of their body odour.[10] It is well known how bees and many flowers are adversely affected by this.

It would indeed be inappropriate for our time to hold up as an ideal for humanity a nature such as Francis of Assisi. Yet his heart and mind, so full of inner warmth and depth, reveal one of the noblest sides of human nature which—if only in humble measure—every plant grower and animal breeder should try to develop.

Certainly the success of beekeeping in the future will depend less on the beekeepers being 'modern' in whatever sense their time dictates, and more on whether they can establish, out of their *whole* humanity, a loving relationship to the being of the bee.

The situation unfortunately is that the more subtle aspects of bee life which one can really only begin to sense through a warm, heart-oriented relationship to the bees, are almost or completely absent from the literature, so that those beekeepers who by nature are not able to establish a contact with the bees beyond just a subjective pleasure, do not have the chance to develop this through study, unless—and this would be something new—someone with a comprehensive knowledge were to reveal the deeper secrets of the bees, thus opening up for others a spiritual access to the deeper realities of hive life.

What prevents the human being from living emotionally and mentally in close contact with the surrounding world is the presence in his being of forces which close him off. Many forces of lower selfhood separate him off in a particular way from his 'interwoven-ness' with the the rest of the world. On the one hand, these forces help him come to a certain self-awareness; on the other, they cause a separation of consciousness from the outer world such as fosters the materialistic deception that the sense world is founded in itself.

More than is generally realized, the question of how to cognise reality is a matter of the higher Self overcoming the lower self in thinking, feeling and willing. For anyone who manages more and more to banish egoism from these three soul forces through self-development—Goethe's *Theory of Colour* and his *Metamorphosis* provide a good schooling for thinking—and reaches the stage where the higher Self is experienced as a spiritual being founded in itself and initially derived only from itself, it becomes clear that the outer world of the senses is likewise founded in the spirit.

Such a path of knowledge is the anthroposophically-oriented spiritual science of Rudolf Steiner. This has made it possible for the human being to penetrate through to the spiritual content of the world in an exact manner and attain a genuine communion with reality. For beekeeping, a conceptual foundation born out of the spirit of anthroposophy is therefore eminently suited to initiating the essential, real, inner spiritual contact between beekeeper and bees, as well as showing the way to healthy techniques of beekeeping.

The author became acquainted more than ten years ago with copies of lectures on bees and ants, given my Rudolf Steiner in 1923. A few years later he gained access to the more or less comprehensive notes of lectures from 1905/08 which, among other things, also discuss bees. He then decided to test the findings in these lectures for their practical applicability and where appropriate to use them as a spiritual-scientific underpinning for beekeeping. This booklet is the the result of those many years of endeavour.

While working on this it became clear that his work would be something of an exception in beekeeping literature. He therefore endeavoured to raise for himself beforehand all possible objections that expert criticism might raise.

Some of the perceptions presented in this booklet are not easy to put into words. During the course of the years many complete or partial re-workings have been required in order to attain sufficient exactness in the presentation. And yet what has been achieved still lags far behind what was envisaged. Further improvements must be reserved for future editions; the author has reasons to wait no longer with the publication. Furthermore a planned sequel of this publication will provide the opportunity to clarify much that has been said here, to expand it, and to address possible misunderstandings.

Iwer Thor Lorenzen

Introduction

When someone sets up an apiary, they usually do it for utilitarian reasons—to pollinate the flowers and to harvest honey. The prospective beekeeper rightly hopes that after an initial learning period, a reasonable profitability will ensue. As justified as the idea of profit may be, it should not be the *a priori* determining factor for the methods employed, for it very easily leads to exploitation. Put clearly: profitability should not be the goal that is sought by cleverly thinking up artificial means to directly bypass the inherent conditions of the bees' existence, but rather, profitability is something that results naturally from appropriate practices of bee husbandry, i.e. those that are derived from the essential nature of the honeybee itself. Beekeeping can only prosper in the long run if it moves in a direction that is determined by the nature of the bee.

There is barely any opposition to this fundamental principle among beekeepers as long as it remains abstract and theoretical, for most beekeepers consider this is how they work. Unfortunately their view of the essential nature of the bee is consistently too superficial for there to be any question of appropriate practices, for it is evident again and again that on essential points only a few have an inkling, let alone a clear consciousness, of the boundaries that are set on beekeeping practices by the nature of the bee. For this reason, the author sees it as a need of our time to present the genuine underlying principles of beekeeping in order to promote healthy practices.

If we ask how beekeeping is possible at all, we come to the conclusion that two *biological* facts are of decisive importance—firstly, that the honeybee has a very special relationship to the floral kingdom, and secondly, that it lives socially in 'colonies'.

Beekeeping would not be possible if the relation of the

honeybee to the flower kingdom were not of a very different nature from that of the wasps (*Vespidae*) and digger wasps (*Sphecidae*), and if the colony's individual bees were not bound together by closer bonds than is the case among the living-, sleeping- or overwintering-communities of wild solitary bees.

We will therefore need to show in this work:

1. the nature of the relation of the bee to the floral world, or put in another way: what is the nature of the flower-feeder wasps and the honey-making bees?
2. What is it that holds a bee colony together?
3. What possibilities emerge for the beekeeper as a result?

Relation of the honeybee to the floral kingdom and the developmental level of digestion in the bee

There are among the insects many feeding specialists. A not insignificant number seek their nourishment from flowers, which offer them pollen and nectar.

The extent to which these insects are attracted to flowers is not at all the same for all groups. Many visit flowers only occasionally, as 'snackers' as it were; others, who as larvae eat roots, leaves, flesh, wool etc., feed as adults (imagos) exclusively on flowers (pollen, nectar and related substances). Wasps and butterflies in general belong to this latter group. We find the highest level of anthophilia (love of flowers) among the bees, comprising not only the honeybee but also thousands of wild bee species. With the bees it is not only the adult that lives exclusively from pollen and nectar, but also the larvae—indeed, one of the criteria determining whether an insect is a bee or a wasp, is the food source of the larvae.

Remarkably, the fact that some insect groups on reaching adulthood seek an entirely different source of food, whereas among the bees this exclusivity exists right from the start, is today barely felt to be a significant or complex question. Superficially one could be tempted to maintain that it is taste that determines the insects in question to choose food from flowers. But this would not explain why a wasp that chooses flower sap and pollen for itself, would not feed its brood with the same. Instead, it considers meat the appropriate food. The subjective measure of liking or disliking breaks down here or is insufficient when one tries to understand the behaviour of the animal. Since there are entire genera, families and even orders that behave in this way, it is clear that subjective motives (which, even if the

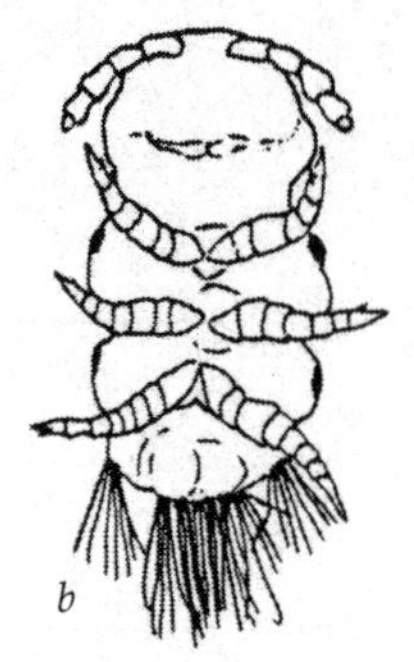

Fig. 1 Early stages of diplopod millepedes:
a) first larval stage of Polyxenus (Linnaeus)
b) newly emerged Julus larva
c) juvenile Julus. The position of the first pair of legs on segments 2–4 is clearly visible. (From an illustration in Kükenthals Handbuch der Zoologie.*)*

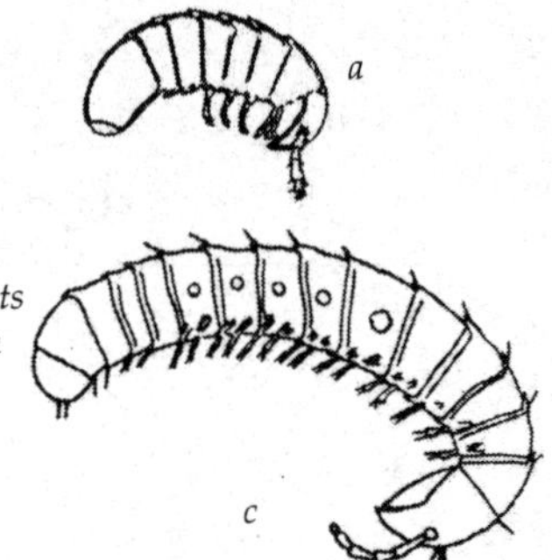

individual animal could be considered an evaluative subject, would only be applicable to that individual) must be ruled out.

In this case the individual animal has no choice as to whether it wants to eat this or that; its life is programmed here as among other things by the characteristics of the species. *The exclusive feeding on flowers by the adults of this insect group arises from the inner imperatives of the organism. Our initial task therefore will be to examine the nature of these imperatives that determine flower-feeding (florivory).*

Among the near relatives of the insects, the millipedes, there is a remarkable phenomenon. The larva of the genus Julus, to take one example, in its early stages closely resembles an insect larva (Fig. 1). While the adults have the many legs to which they owe their name, the young larvae have a pair of legs only on each of the first three segments behind the head, corresponding to the thorax of the insects. The subsequent segments are legless and in this undeveloped state can easily be taken at first sight for the abdominal part of an insect. As the Julus larva grows, the individual segments do not only increase in size, but the posterior end begins in plant-like fashion to 'bud' or proliferate [caudal proliferation]: innumerable segments sprout forth, each of which, with the exception of the last, grows a pair of legs.

A sprouting of this kind does not occur among the insects. Insects emerge with a fixed, species-specific number of segments. (Though in a few cases, e.g. the primitive protura group, the adult insect has a few more segments than the larva, this addition can hardly be seen as a 'sprouting' or proliferation.)

During the larval stage the segments of the body only increase in size. At completion of the larval stage this growth also ceases. Even with plenty of food the adult insect no longer grows.

This incremental reduction of growth forces in the course of insect development has a correlative in the area of the regenerative power of injured animals. Here also we see a diminishment going hand in hand with more complex development. An earthworm, cut in two, will grow a new tail and survive; an adult crab can restore a lost pincer in the course of moulting. By contrast, an adult insect cannot regrow a lost limb; only in the production of whole offspring can it still express the totality of its forces.

When these facts are seen in the right light, it immediately becomes clear out of what necessity adult bees, wasps and butterflies are flower-feeders.

A phenomenon like the caudal proliferation we find in millipedes (anamery), shows that the great surplus of growth forces that enables lower animals to overcome serious injuries without difficulty, is still present here, albeit to a lesser degree. The insects have even less of this surplus, for their number of segments is fixed and, apart from a few unimportant exceptions, already present from the moment the larvae emerge, the segments still only growing in size.

The periodic moults during the larval stage show that the living formative forces unfold their activity rhythmically. After a period of various duration, the forces withdraw from what they have constructed—in particular from the peripheral parts—in order after some time to engage anew.

While it is still possible in connection with the lower animals to talk of a certain detachment or freedom of the growth forces in relation to the organs and structures of the body, the development of insects demonstrates only a partial looseness, allowing these forces, in definite periodicities, to withdraw and re-engage, to partially detach and rebind themselves.

Even this looser relationship comes to an end with the passage through the pupal stage. In the imagos the vegetative forces are bound to the organs or to purely internal functions to such a degree that the metabolism survives only in the processes of nutrition.

The special characteristic we find in the nutrition of adult bees, wasps and butterflies—namely their feeding on pollen and nectar or nothing at all (in this last case they consume reserves from the larval stage)—rests on the fact that, after the transformation from pupa to imago, the organic forces are so exhausted or otherwise tied up that a metabolic weakness results in the form of a reduction in the power of digestion.

That insect flower-feeding is based on a weakness of digestion becomes evident as soon as sufficient value is given to the degree that pollen and nectar are easily digested substances. However, before going further in this direction, we need to look more closely at the nature of digestion itself so as to be able to form an idea as to what is meant by 'weak digestion'.

In order to gain an understanding of the digestive process, we need to keep a certain distance from the notion that we are dealing here with a purely positive process by which the body merely obtains matter that is required for structure and functioning. The existence of an organism does not depend first and foremost on the ingestion of substances, but rather on being able to unfold activity. The *work* on nutrition is more important than its value as mere building material. Seen as a sum of activities, the process of digestion can be differentiated into several phases. The first of these has a wholly negative character.

Food, in the form in which it is ingested, is initially something entirely foreign to the organism; its substance is either a part of inanimate outer nature, or is organized according to the vital organic pattern of another organism, and must therefore first be divested of its foreignness. If it is a mineral substance, then only a breaking-down is necessary; but in the case of animal and vegetable matter, the substance must additionally be stripped of the organic characteristics of the original organism. Substances that are not overcome in this way and which consequently carry their own signature deeper into the organism, cause disorder there. (Think of the poisonous reaction to the protein of another species, even to 'innocent' milk, if this enters directly into the blood.)

Easily digested foods are therefore those that can be broken down easily and those that are deficient in the foreign vital forces and life-energies that have to be killed off. Animals with a weak digestion are those that are able only to a limited degree to 'overpower' and breakdown the food and so become master of the formative forces that are thus released. The latter is the case with butterflies and bees, the former, as will be explored below, applies to nourishment from flowers.

As the plant emerges from the seed and unfolds its growth, it gives its organs—root, shoot, and blossom—not only distinct forms, but also develops a range of forces which can be different and even contrary at the different levels. The more one is able to study Goethe's observations on the metamorphosis of the plant the better can one understand this side of the plant's being. One can then say: of all the parts of the plant, it is the green leaf that best expresses the plant's essential nature. The plant is dynamically (as ideal potential) contained in the leaf with its node; all other organs are variations of this basic organ. 'Forwards and backwards the plant is leaf' (Goethe).

From the insight that leaf, sepal, petal and stamen are all [variations of] the same organ, and that among these the flower-parts must be seen as something secondary and derivative since they do not [photosynthesize or] assimilate, it follows that in the realm of the flower only a fraction of that active process is unfolded which is carried out in full in the leaf.

In the leaf and other green parts—pointing here to an archetypal phenomenon—we see that starch is not only first created but also converted into cellulose, and when necessary converted again back to sugar. In this transformation of carbohydrates, the conversion of starch to cellulose and woody lignin clearly shows a developmental direction towards hardening and lifelessness. This mineralizing tendency is a characteristic of the root region, and is also evident in the high ash content of root wood. The conversion of starch into sugar which is necessary for the transportation of substance from cell to cell, goes more in the direction of the dynamics developed by the plant predominantly in the flower region.

The flower is a quasi-parasite sitting atop the stem. What the green parts have built up while also liberating oxygen, the flower partially breaks down again or burns up while consuming oxygen. (The pollen even at times shows measurable amounts of warmth.[11]) Another indication of this is the release into the air of flower fragrance, showing also that in contrast to the hardening tendency in the root, in the flower we find a dissolving tendency.

In contrast to leaves, in the flower there is no building up, only breaking down. The vegetative forces are as it were held back. In extreme circumstances, as for example with prolifically blossoming miniature fruit trees, this can lead to the tree 'flowering itself to death'. The consuming tendency of the flower—for growing and flowering are indeed polarities in many respects—has a deadening, or at least strongly muting effect on vital processes. It is in this world of deadening, breaking-down and dissolving forces that pollen and nectar receive their unique configuration.

Pollen, rich in protein, shows little resistance to external influences. Since plant protein must be destroyed during digestion in order to be built up again as species-specific animal protein, it is significant that the bees find in pollen-protein something that readily surrenders its foreign nature, and is therefore easy to digest.

Nectar is primarily a watery solution of various sugars (glucose, sucrose and fructose). Traces of oxalic and malic acid are present, as well as scent and taste substances (essential oils). Ash analyses show a significant potassium content as well as not insignificant amounts of phosphates, sulphuric acid, manganese, calcium and natron.

Nectar ripened to honey in the hive contains predominantly simple sugars, whereas in the nectar as gathered by bees from flowers there can be significant levels of (compound) sucrose. The initial supply source of nectar sugar is reserves of starch in the cells near the nectar-secreting glands, the nectaries. The process of transforming these starch reserves by means of dextrin to sucrose and inverted sugars, clearly moves in the direction of break-down. From the ratio of sucrose to inverted sugars we can see how far this breaking-down process—begun in the flower and then carried further by the bees through fermentation—has advanced.

These reverse conversions of carbohydrates reach in inverted sugar a certain limit beyond which lies complete break-down through oxidation (combustion). The phenomenon that occurs there of evaporation and volatilization into air (CO2) and warmth, is the same as that consistently seen in the fragrance substances. Essential oils are always just on the point of evaporating into these finer states. Their behaviour is representative of the flowering process as an event on the boundary between two different conditions.

Apart from essential oils and carbohydrates, vegetable acids too are affected by the breaking down processes so that the flower process essentially encompasses three retrogressive conversions—of carbohydrates, organic acids, and hydrocarbons. These conversions are not unrelated; organic acids after all are regarded as partially derived from hexoses, i.e. sugary substances.

An important function in the flower's breaking-down and vaporization processes is played by sulphur. The plant's sulphur content is at its greatest at flowering time. Certain medicinal plants (camomile, yarrow) are especially high in it. Yet sulphur, whose presence is detectable in flower tissue, is not of particular importance for the processes in question; rather it is present as the waste product of an already completed activity in which it had functioned in finely dispersed 'dynamic' form as the agent [Träger] of the flowering process.

A physiology of honeybee nutrition focussing on process dynamics must regard nectar as something driven by the dynamics proper to the flower (sulphur process) to a specific limit beyond which lie states and releases of energy of a more rarefied nature. By being secreted from the nectaries, the nectar is as it were preserved from being consumed by the sulphur-borne warmth process. (This is possibly a significant thought for an understanding of the nectar-flow and non-nectar-flow of plants.)

By not going beyond this limit—which is indeed the same limit the digestion is heading towards in its first phase—pollen and nectar prove themselves a suitable food for those insects which, as a result of excessive metamorphosis, have diminished digestive power.

To summarize: When we look at the world of insects, we find among them a number that push metamorphosis to the point where a weakening of digestive capacity ensues. When we look at the world of plants, we find that where they too drive metamorphosis to its highest level—in the flower—a dynamic is developed which balances out the deficiency we see in the insects.

It lies within in the healthy forces of the honeybee to start its digestive process at the point where the plant leaves off in the flower. The metabolic organization of the honeybee is incomplete without the processes of the flower. Only together do they form a totality.

To understand bee maladies which are mainly metabolic in nature, and find ways to cure them, we need always to bear this in mind.

The bees actually only begin digestion at the point where, compared with man, the digestive stream passes through the intestinal villi and into the rhythmic system (circulation). What precedes this must be carried out to a certain degree beforehand on the bees' behalf, whether this be through the breaking-down process in the flower or by the fermentation occurring in the bees' own honey production.

From this point of view, we can also consider the following characteristic. When, having passed through the foregut, the food reaches the midgut (the 'stomach'), it does not initially come into contact with the intestinal wall but arrives first in a compartment consisting of a film of cells separated off from the intestinal wall. It is in this 'peritrophic membrane' — effectively a stomach within the stomach — that the pollen is transformed, its nutrient content being rendered absorbable. The peritrophic membrane also marks the beginning of the region in which the bees' organism opposes the acidic nature of the digestive stream with alkalinity — both the food itself and the glandular secretions mixed in it are more or less acidic. We will return to this later on.

The writer has shown elsewhere that the insect metabolic system, seen dynamically, has a particular 'closeness' to the rhythmic system, principally comprising circulation and respiration. However, the metabolism of bees and other flower-feeders, not only has this closeness to rhythm, but in addition is also closely adapted to respiration, that is, the nutrients must have an inherent tendency that allows them to be easily taken hold of and broken down from the respiratory side. This thought brings us to an essential point of the second phase of digestion, to which we now turn.

Once the liquid nutrients have passed through the walls of the midgut into the blood, they meander with the blood flow through the body. Interestingly, the blood flow moves only partly in closed vessels, namely from the tubular heart (in the dorsal side of the abdomen) to the head. Thereafter it moves back freely, but definitely in a regulated manner, through the body. It is exceptionally important and noteworthy that the stream of nourishment is conducted immediately and directly to the head. Here, next to the salivary glands which secrete the ferment for honey production, are also located the glands that secrete the protein-rich larval jelly for the brood. So we need to take note that whereas the vital transformation of nutrients into species-specific substance takes place in humans in the great glands of the abdomen, in the bees it takes place for the most part in the head.

As the flow of nutrients wanders backward from head to abdomen, it is met on all sides by air breathed through the tracheae. And thus, after the phase of nutrient transformation and building up, there follows the phase of breaking down and excretion.

A food like honey, that stands on the edge of breaking down, is exhaled primarily as warmth and air (carbon dioxide). Other waste products however remain longer in the blood. For the bee's organism to maintain its health, it is of primary importance that the breaking-down process be thorough, and that the blood is again and again relieved of waste products. A complete breaking-down depends essentially on the bee organism being able to master the interplay between fluid blood and airy breath. For this, the metabolism requires stimuli from the nerve-sense system by way of taste. It is therefore essential that, apart from sugars, honey also contain organic acids, essential oils and minerals. These substances do not play much of a role as nutrients, but they stimulate the organism and support its activity. Through iron the blood has its required affinity to oxygen. Other minerals are important for establishing the right alkalinity in the blood. The organic acids and essential oils stimulate the nervous system for digestive activity and support excretion, through the urinary vessels, of waste products which accumulate in the blood during metabolism.

Three factors are required for a healthy functioning of the metabolism: firstly, the nerve-sense system must have the necessary stimulus (through taste) so that the organism does not react with indifference to the food, but responds with full digestive *activity*; secondly, the food must be easily broken down; and thirdly, the waste products must be removed totally.

On this foundation, we can now look at bee illness.

On the Treatment of Bee Diseases

After the foregoing it is hardly surprising to learn that bees easily suffer from diarrhoea. Indeed, dysentery is an illness that often occurs in bees. We can disregard the occasional defecation during a prolonged winter by young bees who emerged late in the autumn, since here it is just a matter of a harmless emergency elimination. There are two main factors that can cause dysentery during the winter rest: upset due to queenlessness, and strong external shock. And also poor food. We will look further at this last case.

In certain regions, the bees bring home a honey that comes not from flowers but from a sweet sap exuded by the leaves of linden trees, current bushes, pines, and by aphids. These so-called honeydew honeys are not as pure as stated by Baron von Ehrenfels over a hundred years ago. The living forces of the foreign organism still adhere to these saps because they have not been through the de-vitalization of the flowering process. In addition—and usually given as the only cause—the digestive residues from the ingestion of honeydew honeys, particularly through the dextrin content, are greater than from flower honey and put a greater strain on the gut. However, from the fact that with good food the faecal sac can be filled to its fullest capacity in winter without dysentery ensuing, we can deduce that the bee's organism is normally well able to prevent fermentation as long as it is not compromised by foreign vital forces which cause or encourage it.

The treatment required to cure the bees in this case must be geared towards removing the impurities—by moving the hive for instance—and to re-establishing a healthy digestion. It is reasonable here to try camomile and yarrow on the bees, which are so efficacious in cases of human stomach and intestinal upset.

Give the bees—at first in small, then in larger amounts—a solution of aromatic honey (buckwheat) with an infusion of camomile and yarrow flowers in alternation. To begin with pour 250 g (1/4 lt) of hot water over perhaps 10 g of dried camomile flowers; allow the tea to draw for five minutes, then pour it out and sweeten with 100 g of honey. Give half each to two colonies. On the following evening repeat but with yarrow. If there is no improvement, add a bit of angelica root to the infusion as a stimulant.

Attempts are usually made to avoid the dangers posed by overwintering on honeydew honeys, by removing all the honey and then feeding the bees on sugar solution.

Sucrose is suitable as bee food to the extent that, due to its mineral character, it has nothing that has to be killed off; but it requires greater efforts to break down. In nectar, the flower has already more or less dealt with the inversion, i.e. breaking down compound sugar into simple sugar. With sucrose, the bee must do all this work alone. In summer, the bee quickly exhausts her energies by producing larval jelly and honey, and is therefore a short-lived creature. It follows that having to invert the sugar in the autumn feed puts such demands on the bee that her period of activity in springtime is thereby cut short.

Attempts have been made to relieve the bees of having to invert sugar, by adding a small amount of tartaric acid to the sugar solution. To make the sugar more like honey it would be more natural to introduce the flower-process to the solution by adding camomile tea. Due to its sulphur content and essential oils, camomile is able to a great extent to have the stimulating fiery effect that the flower process does.

Due to its one-sidedness there is another serious deficiency in sugar solution. Not only is there an absence in it of the constituents that stimulate and fire up the digestion, but it is also deficient in all the things that otherwise support and strengthen the organism in its forces and functions, such as fruit acids, essential oils, phosphoric acid, iron, potassium and calcium.

To see how this deficiency can adversely effect the bees' health, we need to look again at the later phases of the bee's digestion.

The real engagement of the bee's organism with its nourishment occurs in the working-together of breathing and blood-circulation, in other words in the realm of the rhythmic system. For a metabolic nature like that of the bee, with its close orientation to respiration and affinity to rhythmicity, honey is a healthy food not only because it is easily broken down, but also because, through the whole dynamic of its configuration, it stimulates the organism in the excretion of waste products. The health of the bee depends to a high degree on maintaining alkalinity in the blood. The loss of this is threatened, however, if the acids that arise during metabolism are not continually eliminated. An important function is carried out here by the Malpighian tubules, whose activity corresponds to that of the kidneys in higher animals. Among the constituents of honey, it is aroma that fosters excretion through these organs—just as in general essential oils play a large role in blood purification cures. Where aroma is missing, as in the feeding of sugar water, the possibility arises of an over-acidification of the blood, particularly in the presence of other conditions such as, for example, stale air and dampness in the hive during winter.

It shows a healthy feeling when in many places thyme, so rich in essential oil, is added to sugar water to make the solution less 'dull'. But melissa, sage, and aniseed should also be considered.* It would certainly be more advisable to add strongly aromatic honeys (buckwheat).

Compromised blood alkalinity provides a breeding-ground for diseases. The beekeepers who maintain that feeding the bees with sugar water in the autumn encourages nosema, are surely correct. Research into bee diseases should not look only one-

---

* Caution is advised with aniseed, for its attraction can easily lead to the hive being robbed and invaded.

sidedly at the so-called pathogens, but essentially at the wider conditions (disposition, constitution) that render the bees susceptible to parasites in the first place. Parasites after all only develop where the conditions favour them.

The cause of nosema is held to be a single-celled organism (*Nosema apis Zander*), that parasitizes the mucous membrane of the midgut. Nosema can be well-established in a colony without there being any outward signs of it. There are apparently other factors involved when the bees suddenly die in large numbers or appear distressed.

The intestinal wall is both the point of absorption of nutrients, and the boundary between the blood—which should be alkaline—and the more or less acidic region of the digestive tract. As described above, the mucous membrane of the midgut periodically separates off as the peritrophic membrane, and is renewed by regenerative cells. It is noteworthy that the upper cell layers contain deposits of strongly birefringent calcium carbonate,[12] and, which is also noteworthy, these inclusions are most prevalent in worker bees. In bees affected with nosema, they are missing almost entirely. Fräulein Köhler, who made this discovery (*Schweizer Bienenzeitung* 1920, Issue 10), assumes that the function of these calcium particles is to neutralize harmful acids. This may be the case from a purely chemical perspective. If we broaden our view to include physiology, we could regard the absence of inclusions as an expression of a loss of blood alkalinity; for eliminating acids from the blood is clearly only one side of the total activity carried out by the organism to oppose the acid-bearing and acid-producing nutrient stream that pushes forward from the digestive tract into the blood. It would therefore be a task deserving much thanks—because important—to research the source of the (occasional?) alkalinity in the content of the midgut. One could perhaps conjecture that the decaying calcium carbonate inclusions that detach along with the peritrophic membrane play an important part here.

Without doubt, the over-acidification of the blood and the

absence of calcium inclusions in the cells of the midgut, are conditional on each other, and are two sides of a disturbance in the total organism.

It was mentioned above that the alkalinity of the blood is put in jeopardy when the bees, during low levels of activity, sit out the winter in cold, damp, stale air. Experience shows that the bees overwinter very well in straw skeps, the straw's permeability preventing bad air and dampness from occurring. It does the bees no harm if the hive opening is left fully open during winter. With fresh air and sufficient food, being able within limits to feel the changes in temperature outside the hive stimulates healthy activity in the organism.

The same conditions that cause a deterioration in the make-up of the blood hold yet another danger, namely mould on the comb. According to research by Count Vitzthum (*Archiv für Bienenkunde,* Issue 3/4) it can be fairly safely assumed that certain moulds, if they get into the bee's intestine, have a toxic effect that impedes peristalsis, causing the young bees that have eaten the pollen to suffer *constipation.*

It is truly a sorry sight to see young bees in May or June with bloated abdomens and quivering wings crawling by the hundreds around the hive stand. Their faecal sac is crammed with pollen mass. Their inability to empty it indicates a problem in the excretory function of the organism, which has its source presumably in an impairment of the nervous system. This is also indicated by the fact that 'May-sick' bees that crawl around the hive stand, manage to fly and partly recover if they are exposed to strong sunlight. The organism clearly needs a stronger impetus to develop the forces that work into the excretory organs.

Since prevention is better than cure, it is particularly important to ensure that during the increasing brood period in April and May everything is kept away that could impair the bees' life-activities, i.e. they must be warm and have mould-free comb, and should be disturbed as little as possible. Preventative measures naturally also include anything that is suited to raising blood alkalinity. A relevant suggestion by Dr Steiner here, for example, is to cultivate and plant out certain flowers to which the bees can go in April and May. The flowers that come into question are those that over and above the usual nutrient value also have a therapeutic one. A guiding principle in choosing these plants needs to be how much they can strengthen the bee organism in its excretory functions and in its quanta of light. Plants can be sought and chosen which, by the characteristics of their indigenous habitats, reveal a special relationship to light, namely alpine plants and plants from high altitudes in Asia.

Foremost to mention is lady's cushion (*Arabis albida*). Anyone who has stood in April before a bed of this and smelled the strong scent of the white flowers and noticed the mass of bees scrabbling about in them can have no doubt that this flower should be considered. Another plant is *Erica carnea,* an early-blossoming heather and also an alpine. Among the primroses there is *Primula cashmeriana* (drumstick primrose). Also look out for the short calyx varieties of *Pulmonaria* (lungworts). Crocus should also not be missing from the 'bee pharmacy'. In order always to have these flowers available at the right time no matter what the conditions, it would be good to grow at least some of them under glass.

In acute cases of nosema and May-sickness, it is recommended to give the bees small portions of good warmed honey (one part honey to two parts water). Trials should also be carried out to see whether juniper berry, which works so beneficially as a diuretic in humans, could not also serve the health of the bees. Juniper berries contain an efficacious essential oil, much glucose, and very small amounts of oxalic acid. Oxalic acid—in homoeopathic doses—is particularly important for the life-functions of an organism, and engagement with it is one of the primal functions of the rhythmic system. By way of exploratory trials, prepare a tea-like infusion (two cups of hot water to one spoonful of chopped juniper berries), strain, and sweeten with honey. Where microscopic investigations of dead bees found on the floor during spring cleaning show the presence of nosema, a prophylactic measure would be to add juniper berry tea, as described above, to the feed. It would also be good to consider adding a little of any of the following: angelica root to stimulate and regulate the digestion; camomille or yarrow to support the metabolism; lemon balm to stimulate the regeneration of the midgut in its periodic renewal; and some horsetail (*Equisetum arvense*) to encourage urination. Preparatory trials would have to establish whether and in what dosage the bees voluntarily ingest and best tolerate the ingredients. Once the optimal dose is established, different combinations could be used.

The writer is unfortunately not currently in a position to carry out the trials himself with sufficient exactness. Since up to now a rational framework for the medical treatment of bees has been wholly lacking, he believes that, for a practical evaluation of the insights given above into the unique characteristics of the bees' metabolic organization, he cannot wait for outer circumstances to make this possible for him personally. It is his firm conviction that the physiology of bee nutritional dynamics presented here, no matter how imperfect perhaps in certain details, nevertheless accords with reality—in other words, it is to be found within the actual tendencies of the most essential nature of the bee and of metabolism in their truest sense. We must wait and see whether the indications for the medical treatment of bee diseases prove correct or not. It is in the nature of the subject that at first only guidelines can be given whose further elaboration requires an intimate knowledge of the healing properties of the plant world. For that we need a physiology of the dynamic forces in the plant. Such a thing unfortunately does not at present exist.

In any case we need to be clear that if nosema is encouraged by feeding the bees sugar solution in the autumn, then the greatest dangers threaten beekeeping from this quarter, just as a general failure to recognize the dynamic link between bee and flower—and therefore also the level of the bee's digestion—could be fatal.

Apart from the above-mentioned diseases, there is also the much feared—because contagious—foulbrood. Whereas with the diseases of adult bees dealt with so far we had to pay particular attention to the fact that bees have a metabolism that is specifically attuned to respiration, with diseases of the brood we need to note that the queen has to digest enormous amounts of protein to enable her to lay the great quantities of eggs. Indeed, at certain times the weight of eggs laid in one day is greater than the queen's own body weight. The bees feeding the queen certainly relieve her of some of this work by feeding her pre-digested jelly; but the queen must carry out the actual protein transformation herself.

The protein in royal jelly does not go straight to the ovaries and then reappear as eggs; it must first be broken down to a certain extent and then built up again as species-specific protein. This protein transformation, which also occurs on a larger scale in the development of the larvae, finishes with the protein being used either as fuel or as building material. If the protein transformation goes wrong, it becomes fertile ground for certain bacilli thus creating the conditions for thoroughly malignant and contagious diseases.

In foulbrood we have a phenomenon in which the primary cause is presumably to be sought in a disturbed protein metabolism in the queen, and where *Bacillus larvae* is involved secondarily. The brood emerging from flawed eggs is reduced to a rubbery, stringy mass smelling like glue and ending up as a dry, tongue-shaped scale in the bottom of the cells.

If this conjecture is correct, then an important measure against foulbrood—apart from disinfection procedures—would be to replace the queen, providing the situation does not necessitate destroying the whole affected colony.

Here we conclude our initial introduction to what, from a sufficiently deep and appreciative look at the relationship of the bee to the flower world, presents itself as a guideline for a therapeutic way of thinking, based on a dynamics-oriented physiology of nutrition.

The Origin of the Honeybee

Views about the origin of an animal must necessarily differ according to one's idea of evolution in general and of the nature of a certain animal in particular.

Darwinism, whose position has dominated almost without exception the literature on the origin of the honeybee, takes into account primarily the physical organism and believes it has attained its level of sophistication gradually through many small steps. Thus in the honeybee particular attention is given, apart from wing venation, to the arrangement of the collecting apparatus which is considered determinative, since these organs are found in varying degrees of development in the 'relatives' of the honeybee and are typical for bee-like insects in general. For there are not only the colony-dwelling honeybees, but also around 20,000 different species of wild bees which in general are solitary.

The whole *Apidae* family which includes all bees both social and solitary, on closer inspection shows a great diversity in the form of the mouth-parts used to collect nectar, and in the hairs and hair structures used to gather pollen, so that with regard to the organizational level of the collecting apparatus there is indeed a broad scale of sophistication. Whether looking more at the sucking organs of the mouth or the collecting apparatus on the hindlegs, the honeybee has an exceptional position in either case, and with the level of its communal life it supersedes them all. At the very bottom we find almost hairless, partially parasitic bees, whose outward appearance differs little from another insect group, the digger wasps. There is a significant biological difference however: whereas amongst even the most primitive bees the larvae are fed exclusively on pollen and nectar, the digger wasps feed their brood on meat (flies and beetles).

Darwinists are of the opinion that bees originated from digger wasps. An offshoot of these wasps is supposed to have weaned itself off meat and then in a long evolution, by degrees and with varying success, perfected the collecting organs. Here too the development of the bee colony is pictured as having begun from the primitive forms of incidental communal life of wild solitary bees (communal places for overwintering and sleeping, 'hall-way' neighbours). By the continuous addition of new elements of communal activity within a kin group (comprising a mother and her offspring), and through division of labour, what is usually called a bee hive or colony is supposed to have arisen.

The insect world itself offers models for a gradual development of the kind assumed by the Darwinist for the phylogeny of the honeybee, namely in the pattern of post-embryonic development of insects with incomplete metamorphosis. The larvae of true bugs [*Heteroptera*], crickets and grasshoppers [*Orthoptera*] when they first emerge already have a certain resemblance to the adult animal, and come closer to this final form with every moult. However, it is open to question whether the ontogeneric mode of development in lower insects can or ought to be transferred to the phylogeny of higher insects, or whether we need rather to reckon with the possibility that in the higher insects the development of the species [phylogeny] occurred, as with the development its individuals [ontogeny], in metamorphoses on a grand scale. According to the spiritual-scientific research of Rudolf Steiner, for the honeybee this is indeed the case.

The mental position on which Darwinism is founded is certainly too passive to cognitively master metamorphoses of such magnitude. For this, in accordance with Goethe, one needs to strive 'to make oneself worthy to participate in the productions of an ever-creating Nature'.

In the previous chapter we attempted to characterize, from one essential angle, the relationship of the honeybee to the floral kingdom. The idea put forward there was that the metabolic organization of the bee is bound up with the dynamics at work in the flower, and only these two working together comprise the whole. This is of great significance and will need to play an important role in a future biology of flowers.

Now the honeybee—as distinct from the butterflies, for whom what has just been said is also valid—does not only feed on flowers, but also makes honey. Both bees and flowers are involved in honey production. The flower produces nectar, and the bee processes it further. In deference to this fact, Linnaeus changed his original Latin designation of *Apis mellifera* (honey-bearer) to *Apis mellifica* (honey-maker). That something essential happens with the nectar when the bee works on it is seen from the fact that bumblebee honey tastes quite different from honeybee honey. Even different honeybee types produce different honey from the same nectar flow. According to Zander the heather honey produced by Caucasian bees is not as viscous as the native bees' honey.

Seen from an outer perspective, the bees' part in honey production consists in adding protein-like nitrogen compounds (enzymes) to the nectar. What the flowers do on the one hand and the bees on the other, does not look so different at first glance. For the flower only appears to be the producer of nectar. In reality it is the green parts of the plant through their up-building activity that create the prerequisite materials. What originates in the shoot is worked on in the flower by the particular dynamics at work there whose breaking-down tendency we saw in the previous chapter, and only here gains the special configuration of nectar. The bees' contribution is to continue and intensify what was begun in the flower, for their enzymes also cause chemical changes tending towards break-down. Different taste qualities are also added by glandular secretions. From a viewpoint of dynamics, it is evident that the bees' part is not only a question of intensification.

The world of forces in honey is the world of forces involved in producing it. The nature of the energies inherent in honey can be seen in its effect on man. Its stimulating and firing-up effect on the metabolic organism has already been mentioned. In the sphere of the rhythmic system it has a favourable effect on the interplay between breath and blood-circulation, being able to counteract disturbances—whether connected with chills or anaemia—at the boundaries where the two systems meet. This effect in man brings to expression the very quality in honey that makes it suitable food for creatures like the flower-feeding insects that have a respiration-oriented metabolism. Honey also gives man healthy structuring forces. Taken regularly, even in just small amounts, it acts as a preventative against sclerosis in the aged and rickets in their offspring.

This configuration of forces comes about thanks to both flower and bee, despite the fact that there is a significant difference in the way each makes its contribution. The production of nectar in the flower is connected to definite external factors, for all plant activity is embedded in a number of earthly and cosmic conditions. The plant is as it were the arena in which nectar arises when certain conditions are present of warmth, light, moisture and soil. Nectar is ultimately the result of the working of cosmic forces in the earth, for the earth on its own cannot bring forth the plant kingdom. The bee is not so dependent on the surrounding world; she fulfils her part out of herself. She can do this because she carries within her, in individualized form, what the plant has to receive from outside. We need to recognize in the honeybee an insect able to continue the process of producing honey from the point at which the flower leaves off, because she has *in* her organism the same forces that influence the plant from *outside* when it makes nectar. It is this aspect of the bee's nature that makes it necessary already at the larval stage that it lives exclusively from flower food.

A perception of this kind of the honeybee as honey-maker requires a different perspective on its origin from that provided by Darwinism, for it is hard to see how anything substantial for the development of the *mellifica* nature of the honeybee is supposed to have occurred purely from a branch of the digger wasps making a transition to a vegetarian mode of life. By contrast the answer that spiritual science must give to the question of the honeybee's origin becomes increasingly satisfying the more closely related to reality the hard-won perspectives on this nature are.

In a lecture in autumn 1923, Rudolf Steiner stated that the honeybee originated from another member of the *Hymenoptera* whose immediate descendants even today live in the closest possible relationship with the flower world, so that only together do insect and flower form a biological unit. The same interconnection of forces that is of such primary importance for bee husbandry, as discussed above, is present here too in a form that could be used right from their inception in both bee husbandry and fig cultivation.

According to Rudolf Steiner's research the honeybee evolved from the fig wasp. Indeed a very similar relationship exists between fig wasp and fig as that between bee and flower. Biologically, fig wasps could justifiably be called 'flower creatures' or 'flower wasps'. In comparison with the honeybee there are indeed considerable differences in form, which could possibly cause doubts among classifiers as to the correctness of the statement. But the possible objections do not weigh so heavily, for after all the difference in the bee between larva and imago is greater than that between fig wasp and honeybee. If the theory of evolution is to make any sense at all, we need to assume for the past a greater capacity and possibility for transformation—in our case in the tertiary era—than applies today. More important for evolution than the fixed form are the forces that pulse in the living blood and are active as soul forces in the nexus of relationships to the outer world, in other words, the physiological and psychological facts as understood in their essential nature.

In the cited lecture, Rudolf Steiner gives a broad description of the nature of the fig wasp, highlighting the attributes that would make possible its transformation into something with the qualities of the bee. A more specific handling of the subject was not necessary, since he was considering original conditions which were far less differentiated than they are today.

It is no simple matter to make the mental leap from the huge numbers of fig-tree species and varieties and the fig wasps specific to them, to the primal situation that existed when fig cultivation first began. To keep our look at this as concrete as possible, we will address the conditions we find today in the Mediterranean cultivation of *Ficus carica* and its species-specific wasp *Blastophaga grossorum Gravenhorst* [aka *Blastophaga psenes*].[13]

This tiny fig wasp, like some other species, is 2 mm long, with very different forms in male and female. The wingless males (Fig. 2b) have a yellow-brown appearance, whereas the females (Fig. 2a) are shiny black on head, thorax and upper side of the abdomen. These creatures spend almost their entire lives in the multiple flower and fruit of the fig. This rather unusual living space begs a closer look.

Fig-trees are characterized by their exceptional vitality. For many centuries they have been propagated by cuttings without any signs of degeneration. Their powerful life force also shows in their ability to create new forms, for the genus *Ficus* has the most species among the flowering plants. More than 600 good species are known, and the most cultivated one, *Ficus carica* [the common fig], has an enormous wealth of cultivars. Some of them crop

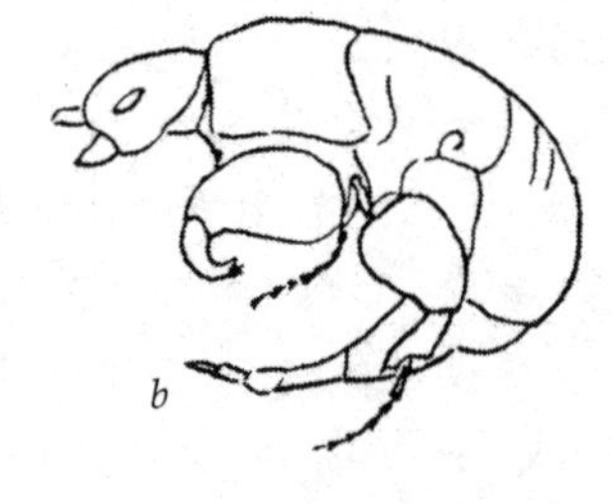

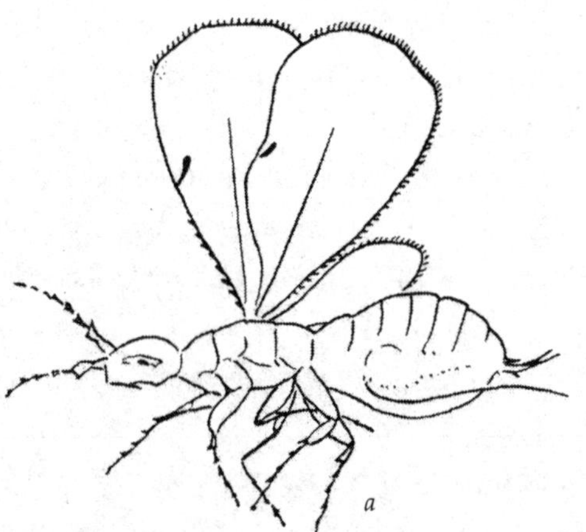

Fig. 2 Female (a) and male (b) of the fig-wasp Blastophaga grossorum Grav. greatly enlarged.

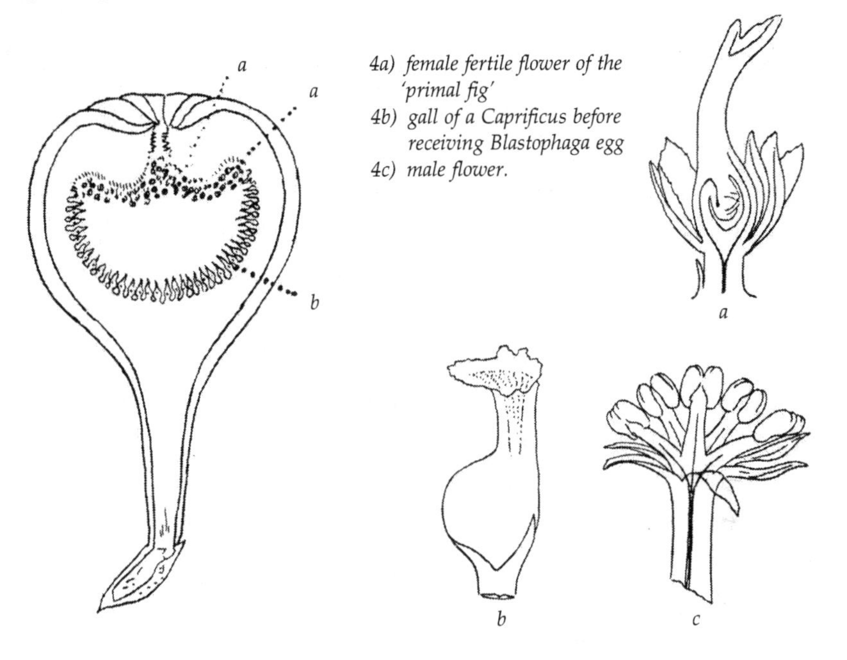

Fig. 3 Longitudinal section of a fig.

Fig. 4 Different flower forms of the fig:

4a) female fertile flower of the 'primal fig'
4b) gall of a Caprificus before receiving Blastophaga egg
4c) male flower.

three times a year, i.e. the tree flowers uninterruptedly throughout the summer, but more strongly at certain times resulting in spring figs, summer figs and autumn/winter figs. Different cultivars yield their main crop at different times, but there are also varieties that crop only once or twice.

What is referred to as the 'fig', is a [flower cluster or] inflorescence [called a 'syconium'], like a pear-shaped pitcher or urn (Fig. 3), which contains different types of flower. We can get an idea of the inner structure of the fig, which can be understood as a metamorphosed twig, if we picture a composite flower [like a sunflower], and imagine the edge of its green underside turning up until it forms a hollow or chamber enclosing the flower, with a tiny opening. The simply structured male flowers (Fig. 4c) are usually situated close to and around the opening (ostiole) (Fig. 3a); the female flowers (Fig. 4a) occupy the back of the chamber (Fig. 3b). However, in some species they are evenly interspersed. Apart from the regular fertile female flowers there are others with shorter styles and shrivelled stigmas (Fig. 4b) in whose ovaries the fig wasp lays her eggs resulting in the swelling of a gall.

The arrangement of these three kinds of fig flower varies according to variety but also according to the cropping cycle within the same variety. We can assume that the conditions most closely resembling the original situation can be found in the common fig (*Ficus carica L Domestica*), referred to by Ravasini as the 'primal fig' *Ficus carica L Eriosyce,* when grown from seed. The arrangement there is as follows. The early figs (profichi) consist of two thirds gall flowers (Fig. 3b) and a third male flowers (Fig. 3a) thus emphasizing the male characteristics through the absence of fertile female flowers; and indeed they produce only pollen-bearing flowers and pollinating insects. The summer figs (fichi) by contrast contain only fertile female flowers. The figs that develop in autumn and produce an overwintering third crop (mamme) contain only gall flowers.

What is united in Ravasini's 'primal fig' is separated in the cultivated varieties, at least in Italy, so that the so-called *Caprificus,* usually grown only in small numbers, embodies the male element, and the *Domestica* the female. In other words, *Caprificus* produces inflorescences along the lines of the early crop (profichi) of the primal fig, and the *Domestica* those along the lines of

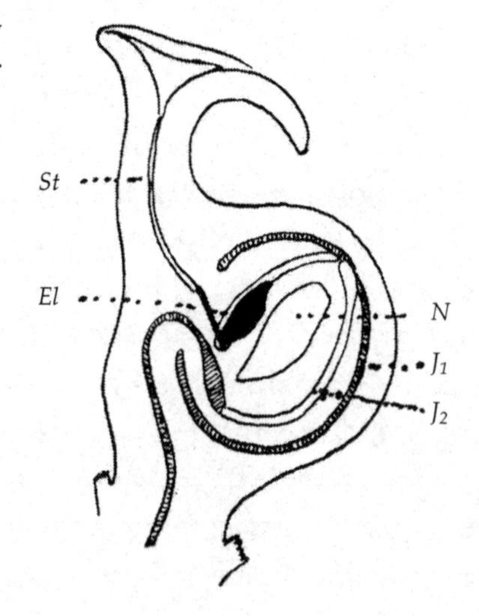

Fig. 5. Vertical section of an ovule containing an egg, from a 'profico' (from Solms-Laubach). N = ovule nucellus with embyo sac. J1 and J2 the two integuments enclosing the same. Egg = Blastophaga egg. Ca = canal through which the egg is introduced.

its second crop. The enormous creativity and flexibility of its nature even allows it at times to develop in the male *Caprificus* proper fertile female flowers among the gall flowers, which actually produce seeds.

The life cycle of the fig wasp is as follows. In April the female pushes her way through the tiny opening into the developing profichi of *Caprificus*, the 'primal fig', and using her ovipositor deposits a single egg in the ovule of each of the gall flowers (Fig. 5) which we have seen to be modified, infertile female flowers. About a month later the larva emerges, which is without feet or trachea (Fig. 6). Full development takes approximately three months. The males are the first to appear. Using their strong three-toothed mandibles they chew through the fairly thick wall of the galls. Once free, they crawl awkwardly around the interior of the fig looking for galls containing females. They chew a small hole into these galls through which they fertilize the female. Later when the inedible fig falls, they die.

The females, using a saw-like structure on their mandibles, enlarge the hole made by the males, leave the gall, linger for a while in the fig and search for a way out. While crawling around, and on passing through the tiny opening, they come into contact with the pollen flowers (Fig. 3a) so that they are covered all over with pollen when they escape from the fig to seek the second crop (fichi) which is female in both the primal fig and in *Domestica*. When pushing into the interior of this fig they often have to forfeit their wings. Once inside they search for a place to deposit their eggs, and pollinate the stigmas of the female flowers (Fig. 7) while doing so. However, depositing eggs in the long styles of these flowers is either impossible or the egg is placed on a spot where it develops only partially before coming to a halt. In *Caprificus*, which also has gall flowers in its second crop, development follows the same course as for the first crop.

Among the many fig species there are some that never have gall flowers. In these cases the insects develop in the normal

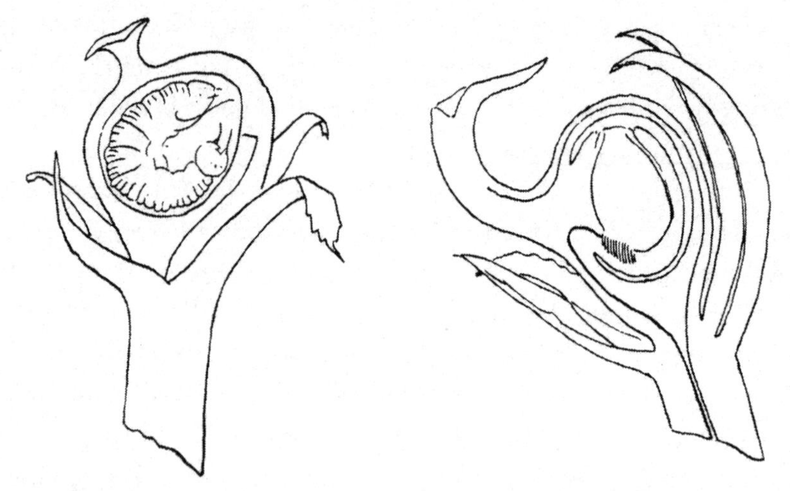

Fig. 6 Gall flower with larva.

Fig. 7 Female fig flower shortly after being fertilized.

female flowers, whose styles are presumably somewhat shorter. In such varied ways the genus *Ficus* demonstrates a youthful flexibility which, considering its great age in the history of the earth, is truly astonishing.

While in the case of *Caprificus* and Ravasini's 'primal fig' there is a third crop, the syconia with gall flowers are impregnated with insect eggs which, if the figs do not fall during the winter, produce insects in the spring. Where insects emerge in late autumn, it might also be a question of overwintering in the open. For, according to Ravasini, *Caprificus* in Italy hardly ever has a third crop, and the 'primal fig', which might also be considered for overwintering, is rare in southern Italy. Yet in spring, the insects appear.

What significance do fig wasps have for fig cultivation? There are fig-tree varieties today that form syconia and attain carpological maturity without any pollination or seed formation. For these varieties, grown in northern Italy, the fig wasps are of no importance. Smyrna figs on the other hand could not survive without them — the fruit falls if it is not pollinated. Fig cultivation in California only became successful once fig wasps were also introduced. Almost all figs traded as dried figs require pollination. Carpologically ripe figs are generally eaten fresh since they do not keep well.

In fact caprifigs also drop prematurely from the tree if they are not occupied by insect eggs. It is indisputable that pollination and impregnation by the insect act as a stimulus to the vital forces. In cases where pollination is not absolutely necessary but takes place nevertheless, the result is an earlier ripening of the fruit.

Because of its importance, fig growers ensure pollination takes place by throwing a number of wasp-occupied caprifigs—usually tied together in a wreath—into the branches of the cultivated fig-trees (*Domestica*) shortly before the wasps are due to emerge. This guarantees that the fig wasps, who are not strong flyers, can reach and pollinate the fichi. In the 'primal fig' the females emerging from the profichi can even make the journey on foot since here the fichi are on the same tree.

The account above of the biological relationship between fig and fig wasp applies to Italy. There is as yet no in-depth research on Greece and the Near East. With the fig-tree's great evolutionary age and its variability and vitality, it is difficult to say what the original conditions might have been. Nevertheless it is telling that *Caprificus* with its more male character—with a few exceptions (e.g. in Algiers)—does not produce edible fruit, and that with Ravasini's 'primal fig' only the second crop, which is female, is edible. It is possible that the improvement of the original fig occurred with a parallel separation of the sexes, whether it be that, as with Ravasini's 'primal fig', the first crop was male and the second female, or that each crop, if there were several, contained both sexes and a part of the female flowers served the insects. Whatever the case, the development of a purely female cultivated variety made it necessary to employ the manipulation of caprification, i.e. hanging wasp-inhabited profichi caprifigs or 'primal figs' in the branches of cultivated fig-trees to ensure pollination.

Considering the scope and depth of biological knowledge

required to initiate such a measure, it does not seem unjustified that significant authorities such as Ravasini and Count zu Solms-Laubach have concluded that fig cultivation could only be the product of an intelligent, mentally alert and learned people.

Expert opinion holds that fig-tree cultivation began in Syria or Palestine, first established by the Semitic peoples. Here indeed were spiritual-mental conditions favourable for this. In the ancient initiation and mystery centres of that region, they had intimate knowledge of the nature of the fig-tree. It was not without grounds that the fig-tree was used as a symbol of initiation. When the ancient literature speaks of 'sitting under the fig-tree', this does not signify, as one often reads, a happy and peaceful state in the sense of ordinary external life, but the experience of a higher level of consciousness, as, for example, the enlightenment of Buddha as he 'sat under the Bodhi tree'. In this elevated state the spiritual-soul aspect of the individual is to a great degree independent of the physical organism. Something similar exists in respect of the fig-tree and its purely organic forces. Where there is lavish growth and adaptability in the plant and animal kingdoms, the formative forces are relatively independent of the the physical structures.

Now, under certain conditions it is definitely possible for someone who has worked towards and attained body-free spiritual vision to be able to study and influence the formative forces of another being (and also, of course, their own—indeed these first of all). In order to understand the measures used when fig cultivation first began, we need to look more closely at the conditions under which fig-wasp development takes place.

As mentioned above, the wasp's insertion and depositing of an egg into the ovule has a similar significance for the further development of the fig as regular fertilization does, namely, it prevents premature dropping of the fruit. But this is not the only effect. Endosperm, ovule and involucre develop in the same way as if fertilization had taken place through the pollen tube. Only the seed-embryo is the exception—its place is substituted by the wasp egg (Figs. 5, 6, 7).

It is certainly highly significant for the inner relationship of fig wasp to fig-tree that both the embryonic and later development of the wasp takes place under the same conditions as the seed formation of the fig. The larva lives from what otherwise would have been used to build up the seed, and it develops under the influence of the same vital formative forces of the fig-tree organism that are active in seed formation, and which later release themselves from the mother plant.

It is exceptionally important for an understanding of the hidden aspects of the fig wasp's nature to give sufficient weight to the fact that it grows *like a surrogate* in the reproductive centre of the fig-tree, being permeated with the same forces as the figs themselves.

If the inner nature of the wild fig-tree had the potential to develop sweetness in its figs, and if the fig wasp likewise possessed the potential to permeate itself permanently with this sweetening-capacity, then it would have to be here [in the reproductive centre] and during the embryonic phase that the sweetening potential would be found.

The more outstanding initiates in the mysteries of Asia Minor would undoubtedly have been able to judge the significance of a creature unfolding its development in the reproductive centre of another organism, especially the fig-tree. They felt called upon to find ways and means to activate the fig-tree's sweetness-potential through the fig wasp—using a dynamic resonance between the two—and to intensify it retroactively.

The solution to this problem was found—reversing the situation with the gall flowers where the seed formation is sacrificed to the insect—by making the insect's development serve a second crop in the fig improvement process. They created conditions which shortened the fig wasp's development period, so that in the same summer the production of a second crop could be introduced.

Similar to the way the prophet Amos set to work, and to the way the fig growers today in Southern Italy, Greece, and the Near East carry out their so-called caprification by hanging profichi caprifigs in the branches of *Domestica* in order to get the wasps to pollinate their figs, so the inaugurator of fig cultivation proceeded in ancient times. He too, according to Steiner, took figs of the primitive form and hung them in the tree to be cultivated. By adopting this measure when the larvae were still at an early stage, conditions were created—by drying—that reduced the length of their development. The insects that were ready earlier due to this procedure now sought to start a second brood. As with the long pistillate flowers of the second crop of the 'primal fig', so also here, in the primitive form, it was not possible to deposit eggs in any way that enabled the larvae to survive beyond a certain point.

This aborted development nevertheless had very real consequences: the forces that would otherwise have enabled a fig wasp to grow and mature now worked into the dynamically related processes of growth and maturation of the figs in such a way that they ripened earlier and became sweet. The whole constitution of the fig-tree even changed over time. Initially these changes were small, but in the course of time they intensified, the characteristics being maintained through vegetative propagation.

We have in the fig growers' caprification methods the practical continuation even today of the measure that, out of deep mystery knowledge, once inaugurated the cultivation of the fig.

But even this action, which took place thousands of years ago, is a faint repetition of an event that occurred, according to Rudolf Steiner, in very distant pre-history. For the honeybee was bred from the fig wasp by wise hands in Atlantis—that part of the earth that is occupied today by the Atlantic Ocean. Outwardly the process was similar to developing the cultivated fig. The endeavour was to reduce the development period of the insect by breeding it in picked and therefore relatively 'dry' figs, and then to release it from its close spatial relationship to the fig-tree, and to gradually extend this inner connection and dynamic bond over the whole floral kingdom. Such metamorphoses were possible at that time.

There were naturally still other aspects involved in the breeding of the honeybee than have been mentioned so far. The author hopes to be able to write about these later in another context. Now, to meet an objection which might with a certain justification be raised, that the bees had existed long before man, we need to note that according to spiritual science man did indeed exist at that time, but not in his present form. The human being is not human by virtue of possessing a physical body with such and such a structure but by being endowed with an individual soul in which the light of the spirit can light up, whether the exterior, as in the tertiary period, had an appearance more reminiscent of an animal or some other appearance, and whether remains of this exist today or not.

The Bee Colony as Individuality and Group-soul

Our knowledge of bee-colony life has seen an extraordinary increase in the last two decades. The results coming to light from large-scale experiments and investigations in numerous research institutions have brought about a change in perception of the nature of the bee colony. Whereas people were formerly disinclined to regard the colony as an 'organism'—and whoever supported this view was in danger of being ridiculed as a 'bee-keeper with an overripe imagination'—the situation has since changed so that the colony is now classified scientifically as 'an organism of the third order'.*

However, referring to the bee colony as an organism has little meaning for an understanding of its inner nature because the concept itself is highly problematic. There is always the danger that it will be taken too abstractly or too superficially, whereas it really needs to be grasped intuitively if it is not to be reduced to a mere term.

The tendency to externalize and nominalize concepts is a particular characteristic of Darwinism under whose influence studies of the bee still very much stand. The fact that it was such eminent specialists as Hermann Müller, v. Buttel-Reepen, and Friese who tried to solve along Darwinian lines the problem of the origin of the honeybee[14] and the evolution that gave rise to its colony,[15] lent such weight to their view that it had a paralysing effect on any other attempt to attain a perception of the nature of the bee based on a deeper level of reality.

The Darwinian view takes the line that the honeybee's communal life had its evolutionary origin in the conditions we find

* More recently classified as a 'superorganism'.

amongst the wild solitary bees, which are thought to have evolved from digger wasps. Since knowledge of the solitary bees' life is not widespread and even in leading beekeeper circles is comparatively unknown, a brief description of it will be given here.[16]

As a rule among the solitary bees each female usually builds her own nest. The location of the nest varies according to genus and species. Numerous species nest in sunny banks, on the edge of sandy pits, in paths, clay walls and wooden posts. Others build their nests on rocks and house ledges, or use hollow stems and snail shells. The simpler constructions have a short vertical or horizontal passage into the earth ending with a widened cell into which nectar and pollen are carried. When sufficient provision has been collected, the female lays an egg on it, and leaves the rest to Mother Nature. As soon as the larva emerges from the egg, it sets about eating the store of food which it gradually consumes until its growth period is complete and pupation ensues. In most cases it emerges from the nest cell in the following year, or sometimes in the year following that, at a time in other words when the mother has long since died. In southern regions some species have two generations per year. A number of bees line the cells with plant fibres, petals and bits of leaves, which in part are cut in a very characteristic species-specific way, before bringing in food and depositing an egg. Very frequently the nests are built next to each other in a row, so that numerous cells fill a single passage or stem. Cells can also branch off from a main passage, or be ordered in pairs.

In certain cases where the development period is only a few months, the mother can still be alive when the adult insects emerge from the cells. This is the case for example with *Halictus sexcinctus* [Six-banded furrow bee], where the mother guards the completed nest. Another bee of this genus, *Halictus quadricinctus* [Giant furrow bee], which even builds a clay comb, also guards the brood nest and is still alive when the offspring emerge.

At favourable sites many nests of the same bee species can

often be found side by side in close proximity. Sunny banks, clay walls, wooden posts often host innumerable nests. Friese found the loam walls of a barn near Budapest to be packed full of *Anthophora parietina* [Anthophorid digger bee] nests. There were between 8,000 and 10,000 bees at work there. Friese writes: 'The walls looked as though they had been hit by a mass of bullets . . . When I waved my net at the innumerable bees, a whole swarm of them went for me, which is not usual for these bees; perhaps they had courage in numbers.' Writing about the small carpenter bee (*Ceratina*) he says: 'After leaving their nest cells in August and September they hollow out dry *Rubus* stems [the bramble and raspberry family] on sunny banks, thus preparing dry winter quarters into which they crawl one after the other, usually filling a whole *Rubus* stem of 20–30 cm with up to 30 individuals. They can easily be found in large numbers in this semi-torpid state during the winter.'

When members of the same species come together like this because the features of a nesting site meet a common need rooted in the species itself—when they congregate, in other words, due to outer circumstances—we cannot yet speak of a communal life in a deeper sense, even if a number of females use a shared stem as the 'corridor' for their otherwise separate nest cells, and generally live in peaceful proximity, as observed and described by Lepeletier in the following example: 'There was a vertical hole in a well-trodden garden path. This was surrounded by eight to ten females of a *Panurgus* species [mining bees], laden with pollen. One female flew out without pollen, and another immediately flew in laden with pollen, unloaded her burden, emerged again and flew away. Many followed in this way. During this time, others arrived with pollen and waited around the edge of the hole until their turn.'[17] What we need to emphasize here as an essential characteristic of the nesting community, is that each animal has the many capabilities of the solitary bees. There is no bodily specialization other than what is purely species- and gender-specific. The common activity arises purely from belonging to the same species which, as we shall see later, is not an abstract concept at all, but a supersensible reality manifesting in physiological and psychological characteristics, a higher unity expressing itself as a 'group-soul'.

In the description of the following bee, collaboration is also carried out while fully preserving the species-specific capabilities of the individual animal.

Adolf Ducke observed a South American *Euglossa* species, in which many females lived in a single nest consisting of over 200 cells arranged in two disc-like layers. The individual cells were made from plant material and earth, and were plastered on the inside with resinous material—were similar in fact to bumblebee cells. A particular characteristic of this bee is the basket on the hind legs, which usually only occurs in social bees. Ducke establishes categorically that the *Euglossa* genus belongs to the solitary bees. 'The difference between social and solitary *Hymenoptera* lies exclusively in the presence of workers among the former with only rudimentary development of genital apparatus.'[18]

As clear as this criterion seems to be, there are nevertheless cases that cannot be ascribed to either group and must be regarded as an intermediate stage.

Although not the richest in species among bee genera, *Halictus* nevertheless has the most varied forms of communal life that are to be found in wild bees. We have already mentioned some notable examples. In the journal *Konowia* (Volume 2, 1923), E. Stöckert describes a group of species which at a certain moment distinguish themselves among the circle of solitary bees by becoming biologically more like bumblebees. It would be beneficial to take a closer look at the biology of this species group, which includes *Halictus malachurus, maculatus, immarginatus, scabiosae* and *calceatus*. Since, apart from small biological differences, these all correspond in the critical points, we can confine ourselves to a description of one species. Stöckert gives a detailed account of *Halictus malachurus,* so this will be chosen here.

Malachurus lives in large numbers and in groups in field tracks and on the sunny edge of woods. In spring one can see many small piles at the nesting sites which arise from the bees making new nests or repairing old ones and pushing up quantities of earth behind them as 'building rubble'. The walls of the nest tunnels are cemented with a type of saliva, and the cell walls are treated with a complete coating of this.

The females that are born in autumn overwinter in this mother construction. In spring usually only one bee remains behind while the others make or look for nests of their own. At the beginning of June the nest sites are quieter until the summer brood emerges towards the end of the month. The females emerge first and are somewhat smaller than the mother bee. (Bee researchers earlier thought these summer females were a different species and classified them as *Hal. longulus*.) Zander's research shows that these summer females as a rule remain unfertilized, so that in circumstances where reproducing becomes necessary—their ovaries are smaller but functional—this can only occur by parthenogenesis. They also do not establish their own nests, but remain with the mother and work alongside her. About a week after they have emerged they begin to forage and supply the nest cells with pellets of food.

The following can briefly be said about the nesting site. The burrow is round, 5 mm wide narrowing to 2½ mm at the entrance, and around 25 cm deep. The nest cells, roughly 9–11 mm long and 5–6 mm wide, are constructed horizontally off this main passage. Once the cell is provisioned with a food pellet and an egg, it is plugged with earth. A bee stands guard at the main entrance to prevent intrusion from enemies or parasites, plugging the entrance with its head, like a lid. When a bee returns, the 'lid' opens, i.e. the guard-bee backs off a little. As soon as the returner has crawled in, the entrance is immediately closed again with the guard's head. If foreign bees or enemies approach, the guard-bee remains at its post and defends itself if necessary. (Aurivillius[19] experimented with this at a nest site near Kronstadt. When he pulled out the guard-bee with a pair of tweezers, a replacement immediately filled its post. He pulled this one out, and another arrived. After this had been repeated four times, the fifth bee set about sealing up the entrance with bits of earth, until just a small hole remained, through which it poked the tip of its abdomen with the sting.) The entrance is also closed with earth in the evenings and on rainy days. The mother herself often stands guard. In August the burrow contains around ten to twelve individuals; as many would have already died since the lifespan of a summer female is around four to six weeks. The mother bees, on the other hand, live for over a year.

The males appear from the end of July onwards. They swarm around the nest sites and die soon after mating. They ignore the summer females and are tolerated if they enter foreign nests. They are most numerous in August and September when the autumn females that will see out the winter appear. At the end of September, beginning of October, when the fertilization of the future mother bees is accomplished, the *malachurus* 'colony' breaks up. The males leave the nest for good, and the females, among whom are also occasionally some late summer females, retreat to an area of the mother nest which will be their winter quarters, having first thickly plugged up the entrance.

Stöckert summarizes his view on nesting communities of the kind just described for *Hal. malachurus,* as follows: he says that among Halictus 'some species, namely *malachurus, maculatus* and *immarginatus,* probably also *scabiosae* and still other species, have already crossed the threshold to a social life, and together with the honeybee, bumblebees and stingless bees (*Melipona* and *Trigona*) must be regarded as "social" bees'. 'Although the nesting communities of the above *Halictus* species are still very primitive and barely deserve the name of "colony", they nevertheless already exhibit the essential attributes of a truly social insect community, namely, the mother bee living at the same time as her offspring, the extended mutual toleration—even support—and cohabitation of the mother bee with the majority of young females in the mother nest, and finally in particular the development of a kind of "worker caste" from the smaller, normally unfertilized helper females. For according to Reuter, the most characteristic attribute of a truly social insect community is the differentiation of a species into different castes (sexual bees and worker bees).'

Although the writer can readily agree with the view that the *Halictus* species in question 'has already crossed the threshold to a social life', he cannot go as far as to put it in the same category as the honeybee; for there is an essential difference whether fertilization of a number of females fails to take place due to lack of oestrus, or lack of a wish to mate, and yet parthenogenetic reproduction is still possible (the summer females' ovaries, as mentioned above, are not strongly developed but are functional nevertheless), or whether all reproduction is suppressed and even, as we shall see, its forces undergo a metamorphosis into a soul-spiritual element, as is the case in a honeybee colony.

The facts call for a distinction to be made of three categories in the *Apidae*, the bee family:

1. communities on the basis of being the same species;
2. communities in which, in addition, a portion of the females remain unfertilized due to absence of oestrus, but parthenogenetic reproduction remains possible;
3. communities in which reproductive ability is suppressed in all females except one mother bee.

It becomes apparent in this grouping that a high level of social life is connected with a holding back of soul- and organic-forces.

The loss of reproductive ability in all the female individuals of a bee colony, with the exception of the queen, is indeed the essential and, for our understanding, decisive difference between a bee *colony* and a simple bee family or just a congregation of bees of the same species. The high degree of independence enjoyed by solitary bees in possessing all the capacities of their species, does naturally not exist for the members of a bee colony. In the colony the separate categories (worker, queen, drone) are heavily reliant on each other, even though their bodily differentiation for specific functions is not nearly so marked as with some ants where even the workers are differentiated into ranks.

In an overview and appraisal of conditions among solitary and social bees, one thing that needs to be recognized and which highlights the inadequacy of the Darwinian view of the bee colony's evolution (gradual perfecting of the gathering apparatus and increased division of labour among related groups) is the following. The perfecting of the physical organism and the suspension of reproductive ability or suppression of the reproductive instinct among the worker bees does not represent a simple straight evolutionary line. For in the first instance the soul-forces push more strongly into the physical organization causing capacities in the fluid realm of formative processes to solidify, as it were, as 'tools'. In the second case soul-forces (instincts) are prevented from becoming bodily active. Both these lie outside the individual animal's sphere of choice, and can only be understood in relation to the principle of the 'being' of the species—superordinate to all individual bees—and its reality as a group-soul.

Now what do we understand by 'group-soul'?

It is extremely difficult for a modern person to imagine a being whose primary characteristic is wisdom-filled soul-inwardness, let alone the connection of this being to the plurality of hive inhabitants. Our understanding is helped if we always keep in mind that members of the same species can cross-breed without limit, in other words they form a physiological unit based on a parity of formative forces. If we now consider that these species-specific formative forces are the link and intermediary between the organic-bodily nature and the instinctual soul nature of the individual animal, and that the animal's activities cannot go beyond species-specific actions, then we can clarify the connection between the group-soul and the plurality of the species' members in the following way.

Every animal has a physical body of organs, a body of living formative forces (etheric body), and a sentient soul-body which is the bearer of instincts and with good reason can also be called the astral body. In Figure 8 the inner circle symbolizes the physical body, and also the ovum. The etheric formative forces (hatched circle) work from the cosmos into the physical body, whose essential attributes are that it is material and occupies space. Etheric forces are non-material; they work in a non-spatial, suctional manner and permeate the physical body, infusing it with growth. The astral body, consisting of instincts and drives (represented by chevrons), is superordinate to both the first two force systems.

At the essence of its being, the group-soul has an 'I', and thereby possesses the nature of 'personhood'. The 'I' as the centre of inner being is enveloped by an astral and an etheric body (Fig. 9). The forces of these two organisms create a contact between the group-soul and the remaining world. Direct contact, however, is at first only with those realms consisting of processes and realities at the purely spiritual, soul and vital levels. With the material world the contact is only indirect—through the individual members of the species. Certain parts of the etheric and astral organisation protrude like extensions into the earthly-physical world, and organize the embryonic development of the individual animal (Fig. 10). Once development is completed, all individuals remain intimately connected, at a soul level, with the group-'I', whose organs they are.

This applies to all animals in general. Before we look at the particular conditions in the bee colony, we will briefly summarize the essentials of the outer life of the honeybee. For more detailed information, see the works of K. v. Frisch, Zander and Leuenberger.

In the nesting communities of solitary bees there are only male and female animals. By contrast, in a bee colony, alongside the single female known as the queen and who alone is responsible for laying eggs, there is also the mass of female worker bees

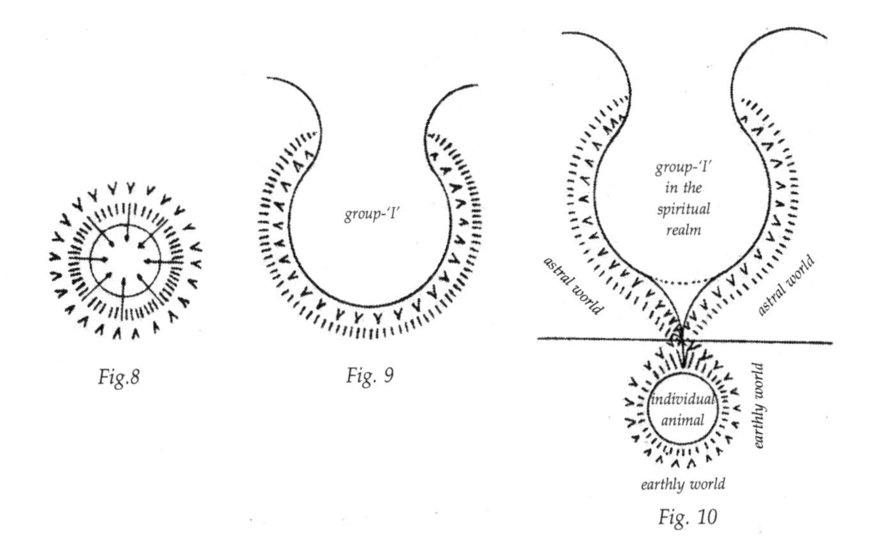

Fig.8

Fig. 9

Fig. 10

whose genital apparatus normally remains rudimentary, and also the male drones.

The queen is larger and slimmer than the worker bees and is distinguished by having strongly developed ovaries and no collecting apparatus on her hind legs. Her proboscis is less developed than in the workers. A queen requires 16 days to grow, and does so in a sack-like cell, with her head facing downwards. Her nourishment consist of highly nutritious royal jelly. Once a young queen is fertilized, she devotes herself almost exclusively during the summer months to egg laying. At the height of the egg laying period (May–June), the daily production is often more than 2000 eggs, whose collective weight is more than the queen herself. This productivity is only possible because the queen is attended during her egg laying by helpers—around twelve bees comprising her 'court'—who constantly feed her with the protein-rich, milk-like juice. Fertilized eggs produce females (queens or workers); unfertilized eggs, which the queen can also lay, produce drones.

Worker bees are characterized by the specialized development of organs which serve for building comb, feeding the brood, and gathering food. In a regular colony, depending on the time of year, they number between 20,000 and 50,000 individuals. Their development from egg to imago takes 21 days.

Three days after the egg is laid, a legless larva emerges which is fed for three days on a jelly produced by young nurse bees. During the following three days it is also fed pollen and honey. At this point the actual growth period is completed; the larva spins itself a silken shroud, and its honeycomb cell is sealed with a thin cap of wax and pollen. In complete seclusion, the mysterious processes now take place that transform the worm-like grub into the winged insect.

On leaving its cell, the young bee soon begins a sequence of activities whose succession and duration are closely bound to the maturation of specific organs. According to *Untersuchungen über die Arbeitsteilung im Bienenstaat* [Research into the Division of Labour in the Bee colony] by Rösch, one to three day-old bees are engaged in cell cleaning; four to six day-olds feed the older larvae with pollen and honey; seven to ten day-olds also feed larvae but with the jelly from their now mature glands. From 12–18 days, when the wax glands begin to function, the bees engage in comb-building. After 18 days they also do guard duty at the flight hole. In abnormal or extraordinary circumstances, for example when there are no young bees, the older bees will revert to the functions of the younger ones, and conversely, when there are no forager bees, younger bees will fly out to forage before their three-week stint of interior duties is complete. When, after a developmental period of two to three weeks, the worker bee becomes a forager, around three more weeks of exterior service follow, and then, in the summer months, its life will be over. Only late brood of August and September live to see the new spring. In the rapid turnover of generations the queen is the still point; she usually lives around three to four years.

The situation for drones is different. They develop in 24 days in larger cells than those for the worker bees. Drones are ungainly, fairly helpless creatures that remain for the most part on the sidelines of the hive, and only fly out with a loud buzzing at midday or early afternoon during fine weather to mate with young queens. The first drones appear in early to mid-May. Under normal conditions there are usually a few hundred of them in the hive. At the end of July or beginning of August, sometimes even earlier, the drones are removed in the so-called 'slaughter of the drones'.

The following can be said of colony behaviour during the course of the year.

On sunny days in early spring when the air temperature is at ten degrees or more, the bees make their first flight out of the hive to relieve themselves of digestive remains, either spraying out the liquid content of the faecal sac during flight, or depositing brown droplets while sitting outside the hive. A sudden drop in temperature or onset of snow during these cleansing flights poses a danger for the bees, for they easily freeze when separated from the colony as a whole. Already in March, when the crocuses and willow catkins are in bloom, fresh pollen and also water are gathered. The queen now begins to lay more eggs, and with increasing forage and a growing number of young bees, the brood nest is quickly extended. Comb-building begins around fruit blossoming time. The preparation of new comb with larger cells for drone brood tells us the time for swarming is approaching. Under favourable conditions a colony is strong enough by the last third of May that swarming can begin. Before this, queen cells are built and around a week before swarming, eggs are laid in them.

The zeal for work abates markedly as soon as the young queens begin to develop. Once the development of the first queen has progressed to, or is approaching, pupation—in other words when the development is halfway complete—the old queen departs with a large portion of the colony. The actual act of swarming is usually preceded by a 'nervous' flying to and fro of numerous bees, who then suddenly disappear again into the hive. Soon afterwards a mass of bees streams from the hive, which makes a very unique impression. The bulk of the bees streams determinedly and in perfect unison in wide columns from the flight board 'like something boiling over'. Others plunge 'head over heels' from the entrance, as though driven by a storm wind. Once the swarm cloud is out, the bees quickly gather into a cluster around the old queen on a spot previously reconnoitered by scout bees.

Primary swarms [also called 'mother swarms' in German], i.e. those containing an old queen, quickly set to foraging and building comb. Larger swarms that are caught in an empty swarm basket in the morning can have a number of hand-sized combs built by the evening of the same day. In good weather and with abundant forage, building work is rapid, so that after a week there is already sufficient comb to develop the colony and overwinter.

Around eight days after the departure of the primary swarm, when the 'piping' and 'quacking' can be heard of the first young queens whose outer development is complete, the first afterswarm occurs. It usually takes longer for this swarm to settle and compose itself. The cluster also often has more than one tip, revealing the presence of a number of queens, and thus having far less of the enclosed quality of the primary swarm. Once the swarm is hived, the superfluous queens are quickly killed; they are usually to be found the next morning lying dead on the ground in front of the hive stand. In the Egyptian Bee (*Apis mellifica fasciata*), which has some biological peculiarities, it can happen that a number of young queens are present in a swarm for weeks. As soon as one is fertilized, however, all the others are disposed of. In indigenous bees the nuptial flight of the young queen of smaller swarms takes place, good weather permitting, after only a few days. Certain new breeding methods, that are not based on the needs of the bee, do indeed often have a delaying effect on this process that is so vital to ensuring the existence of the colony.

Good afterswarms also build industriously and usually make beautiful worker comb whereas primary swarms often produce a surplus of drone cells.

Apart from the primary and first afterswarm, an overwintered colony, when its development is allowed to run its course, as a rule casts a second and third afterswarm at three-day intervals; in addition, where a queen has remained behind, a new colony arises from the remaining bees. An early primary swarm will sometimes swarm a second time when the heather blooms (heather swarm).

This powerful multiplying phenomenon played and still plays an important role in the old heather bee-husbandry. We can leave aside here the question of whether these methods are still economic today; for our task is not primarily about the income of the beekeeper, but the life and behaviour of the bee colony, the determinants of which we are seeking to understand. And here we come to a factor not mentioned hitherto but which expresses very specifically how in a bee colony we have more than a working community based on related individuals or merely a collection of bees with a queen.

It is only relatively recently that we have learnt of the warmth conditions prevailing in a hive.

In 1915 von Buttel-Reepen wrote in his book *Leben und Wesen der Bienen* [Life and Nature of the Bees]: 'During the brood period there is a constant temperature in the brood nest of 32–35°.'

In 1921 Zander says in his work *Leben der Biene* [The Life of the Bee]: 'While there is brood in the hive, the temperature is 34–35°, at least in the brood nest.'

In 1922 Küstenmacher writes in his *Unsere Biene* [Our Bee]: 'If we measure the temperature of a bee colony, we find that in summer and winter it has a temperature a few degrees above 30°C.'

These quotations on the warmth conditions in a beehive have the character of a mere statement of facts. The further discussion in the above works also shows no sign of having any inkling as to the nature of the secret we are getting close to here. Whether the temperature were slightly higher or lower would apparently not be regarded as significant, the readings would probably simply have been noted and understood in terms of heat generated by digestion, and the combined warmth of the many inhabitants of the hive resulting in the temperature at the recorded level. Zander says clearly: 'The warmth in the hive is the individual warmth given off from the bees.'

In 1926 Hess published (Zürich) the results of an exceptionally important experiment.[20] On a hot day when the temperature in the shade was 34°C, he put a heating pad under a hive, causing the temperature in the box to rise to 40°C. Nevertheless, the temperature in the brood nest did not rise above 36°C.

We stand here before the astounding fact that the colony can not only increase the warmth level in the hive during cold external temperatures, but can also decrease it during high temperatures. Thus it is no longer possible to regard the temperature in the hive merely as the warmth given off from the individual bees; for we find here the same conditions we find in the higher animals within the boundaries of their skin.

Hess, who understands that an explanation purely in terms of physics is not possible here, does not hesitate to draw the logical conclusions. He says: 'The bees of a colony in their totality perform a series of activities which give the colony the status of a self-contained entity. This entity—seen as an organism of a higher order—is regulated by joint activities which can be placed at the same level as the vegetative functions of higher organisms. To these, among other things, belongs the regulation of temperature within the colony.'

Careful research into the warmth conditions in the hive has led to the recognition of the bee colony as an organism of a higher order. With all respect for the courage Hess has shown by drawing the above conclusion, which presumably brought him no sympathy from the 'atomists', the writer cannot but regard even this view as insufficient because the concept of 'organism' can already be applied to a purely vegetative plant, and therefore does not adequately address the soul element permeating the animal organism.

We have already mentioned that in the animal kingdom all members of a species and of a proper group (which can often encompass whole genera of abstract taxonomy) belong to a group-soul. For a spiritual researcher these group-souls are not the product of fantasy, but objects of spiritual perception and experience. Using the very concrete indications given by Rudolf Steiner[21] concerning the honeybee and its 'group-soul-ness', the writer would like to make the following remarks:

Unlike other animal groups, honeybees (that is, members of the taxonomic species *Apis mellifica*) do not all have *one* common group-soul, but each colony, at the soul level, is an individuality in its own right; one could also call it a 'bee personality'. The group-soul—the conscious spirit of the bee colony—lives in super-sensible worlds and can be greeted there in full consciousness by the seer as a sister being. The group-souls of the bee colonies are to be included among the higher intelligences holding sway in the cosmos; their consciousness reaches as far as the plane of Providence. In wisdom they foresaw the necessity of linking their development with that of the earth, and acted accordingly. This exceptional greatness of the bee spirits—in certain respects they are higher than the human soul—is connected with the fact that they did not take into themselves the forces of lower egotism—in biblical terms: were not part of the Fall—and therefore were not driven, from a cosmic-spiritual point of view, into the depths that human evolution had to undergo as a result of the influence of lower selfhood. It is fully in the nature of the sublime and selfless purity of the bee colony's being, for whom all feeling of lust is foreign, that where it comes into contact with the sexual component of the physical line of evolution (origin in the fig wasp!), it suspends the reproductive drive in the worker bees of the specific colony, and in the one queen lets it become an innocent, almost plant-like burgeoning (one cannot well call the mass production of eggs anything else). Selfless forces of love expressed as soul warmth, and high wisdom manifesting powerfully in activity, define the group-souls of bee colonies in which the soul forces are under the full rulership of self-consciousness.

Initially it may seem impossible to verify that Rudolf Steiner's indications here result from a real knowledge of actual facts, and yet it is inasmuch as, using the mode and method of this research, a fact was established that was not discovered by outer science until almost 20 years later.

In a lecture in Stuttgart on 14 September 1907, Steiner said, *inter alia,* the following: 'If we investigate the temperature in a bee colony, we find it has roughly the same temperature as human blood. The whole colony maintains this blood temperature because in its inner being it has the same origin as human blood.' Far more of the essential and all-encompassing nature of the subject is contained in this statement than in the formulation represented by Hess on the basis of his experiment. Anyone believing that Steiner's statement must be regarded as fantastical need only consider that through this research into the real relation of the group-soul to the physical bees, a result was attained that is outwardly verifiable and has since been confirmed as correct from other quarters.

The temperature goes below the usual level in the hive during winter, and also if there is no brood during summer, because the connection of the group-soul to the individual bees is looser at these times. However, it just needs a strong blow against the side of the hive to bring the temperature of the bee cluster up to 36°, even in frosty winter. This touches on the question of the incorporation of the group-soul into the bee colony, a proper understanding of which is of fundamental significance for the healthy continuance of beekeeping.

A soul-being that turns a number of bees into a colony with an organized warmth body, and even suppresses and transforms the reproductive forces in the worker bees, can only exert its influence fully on the individual bees if certain conditions are met. We will acquaint ourselves with these factors by looking at the genesis of a colony.

Let us imagine a colony which, having overwintered, casts a number of swarms during May and June. In order for portions to be able to separate off as swarms from the unity of the mother colony, permeated as it is with group-soul forces, a 'chaoticizing' and atomizing must first take place. This process is initiated by the laying of new queen eggs. From this point onwards there is a noticeable slackening of activity in the bees. German beekeepers say that a 'Schwarmdusel' [a swarm doziness] has taken hold of the colony. When eight days later the old queen departs with a large portion of the bees, it is actually the old colony—seen at the soul level—that is leaving. This swarm is therefore technically called [in German] the 'old-swarm' or 'mother-swarm' [or in English, the primary swarm].

As a rule more swarms separate off from the remaining bees—usually on the ninth, twelfth and fifteenth day after the primary swarm. These afterswarms are not yet colonies in a true sense; for at this stage the group-soul takes hold of the individual bees with no more intensity than in the case of *Malachurus* (see above). The bees become an actual colony—the bee colony is born at a soul level—the moment a young queen is fertilized. From this moment onwards not only is the succession of worker bees guaranteed, but it now becomes possible for the asexually constituted group-soul to take hold of the individual bees more intensely, resulting in a complete suppression in them of the sexual element. If for any reason the fertilization does not take place, then the preconditions for the suppression remain absent. The consequence then is that after a while the worker bees, exactly like the summer females of *Malachurus,* start laying unfertilized eggs from which only drones emerge.

The suppression of the reproductive forces in the worker bees of a regular colony does not mean the complete exclusion or even destruction of these forces, but rather brings about a metamorphosis of them. They do not organize new bee bodies, but instead organize the life-body of the totality: they become the life-spirit that manifests itself in the wisdom-filled, egotism-free collaboration of the colony's members. From the group-soul's power to master and spiritualize the organic forces predisposed for reproduction, arises the bee colony's inner cohesion and harmony. The group-'I' lives, in spiritual wakefulness as an ordering consciousness, in the worker-bee forces that have become, or have always remained, free of the body. In a certain respect, the queen is the 'bearer' of this group-'I'; its warm, love-filled inwardness descends into the warmth organism of the individual bees and thus orders and organizes the warmth body of the colony as a unified whole. The bee colony's management of temperature is not just a phenomenon of physics or organic life, but also one of soul.

The development of a colony therefore proceeds inwardly and outwardly in three stages. In the auric, soul-level glow of the eggs and larvae of the developing queens, with its atomizing effect on the mother colony, we see the preparation of a new group-soul towards embodiment (first stage). In the departing swarm (afterswarm) the group-soul holds the bees in a loose association (second stage). Once the queen is fertilized, the group-soul's connection to this association and the worker bees that feel a part of it, gives rise to the enclosed 'self-containedness' that comes to expression in the regulated warmth conditions in the hive, and in the suppression of the reproductive instinct in the worker bees; the unity of feeling (first stage) has now also become a unity of will (third stage).

The functions carried out by the three kinds of bee in colony life have already been discussed. We now turn to the role they play specifically for the group-soul. For this we need to consider a set of facts which the group-soul and the bees have in common, and at the same time forms the soul-organizational connection between them. We can gain an insight into this connection by looking at the development of the bees from a cosmological view point.

In order for a bee to develop in its egg, not only must the known physical conditions be fulfilled (warmth, special food, fertilization—except for drones), but the life-etheric and soul-astral forces specific to the species—i.e. the forces corresponding to the being of the group-soul and forming its 'sheaths'—must also work in from the cosmos. Physical development receives its impulse and direction from supersensible forces working, in a manner determined and selected by the group-soul, out of the creativity of the cosmos. These forces in their totality form the etheric and astral organization of the bees. In their interaction with body-forming substance, they initially generate three dynamic systems: a rhythmic system; a rhythm-oriented metabolic system; and a rhythm-muting, growth-suppressing consciousness system—from which the corresponding outer organs arise.

The completion of the shaping processes in the creation of an individual organ as well as of the whole insect, is the result of the activity of the astral organism. This organism unfolds its formative impulses in connection with certain processes in the macrocosm. The honeybee is particularly bound up with the movement of the sun, the group-soul being 'at home' as it were in the spiritual sphere of the sun. When, at the inception of a young bee, forces work in from the cosmos and set in motion the development of the egg, these forces are determined—by their quality and through a kind of selection—by the group-soul. It is as though a tip of the group-soul is obtruded into the earthly world in conjunction with the ongoing rotation of the sun around its own axis (see Fig. 10). How much of the etheric and astral organism goes into the development of the physical body, or remains 'body-free' during this process, depends on the period of solar-rotation that is required until the organ formation is complete. The developmental period of 21 days for worker bees sets a limit, founded in astral processes (solar rotation), beyond which the forces of the supersensible organisms may no longer engage in bodily development without resulting in a loosening (weakening) of the connection to the group-soul.

That the queen, due to special circumstances, is already complete after only 16 days, means that she has a larger quantum of body-free forces—i.e. forces remaining in the sun realm of the group-souls—at her disposal. This gives her a unique position within the totality of the colony: she becomes its centre of life in both a physical and soul sense. As we have seen above, she can rightly be called the 'bearer' of the group-soul in the colony.

By their longer period of development, the worker bees come very close to the point at which the solar-cosmic element is used up. What nevertheless keeps them closely bound to the group-soul, and imbues the connection with soul quality, are the inhibited reproductive forces that remain dormant. In the worker bees the forces which, were they not held back, would otherwise lead to physiological [i.e. sexual] maturity, remain at the body-free soul level.

The drones, who require 24 days to develop, have the least body-free forces, and consequently also the loosest connection to the group-soul. This can be seen, among other things, in the way they easily rove from hive to hive.

The colony's three types of bee differ in the range and quality of the forces through which, as a result of the length of their development, they remain connected to the group-soul. We do not have a proper idea of this group-soul if we imagine it abstractly as something uniform. Rather it is a threefold being combining in itself a willing, feeling and mental element. And these three components of its being come as soul forces under the complete mastery of its self-consciousness, the group-'I'. The soul forces order the three types of bee for the group-'I' so that the queen is immersed in the will element, and the workers in the feeling element, which from what has been said above is understandable. The drones, who fertilize the queen—the last stage in the genesis of a colony—bring about a higher level of 'colony consciousness'. As the instigators of inner wakefulness they are to be allocated to the brightest soul quality—the mental element. It is not by chance but very much to the point that the old beekeepers of the heather regions called the colonies in which the drones had fulfilled their task, 'clever'.

What brings the three soul forces in the group-soul to a higher unity is the group-'I' presiding over them. There are a number of indicators that show a particular connection of the group-'I' with the queen. If a queen dies, for example, or is removed by the beekeeper, the colony quickly becomes agitated and the reproductive instinct begins to revive in the worker bees. It still takes some time though—around three weeks—before the egg laying begins. In the Egyptian bee, which significantly does not form a winter cluster, the worker bees usually take over the egg-laying work already on the following day. Out of their own nature, the individual bees will develop the reproductive instinct; this can only be suppressed and transformed by the superordinate asexually-oriented group-soul under the influence of the group-'I'. This clarifies in itself 'how the bees know so quickly that the queen is missing'; they sense they have fallen away from the dominion of the group-'I' and that the reviving reproductive instinct is causing physiological changes in them.

In the reverse situation where a new queen is introduced to bees that have lost their old one, the question—of exceptional importance for the practice of beekeeping—arises regarding the conditions under which the group-'I' that is connected with the queen can extend its influence over the bees. This is a question that at the same time sheds significant light on the less evident processes of the third stage of colony genesis. We will take a closer look at this.

In beekeeping literature there is much mention of various scents [or nowadays, pheromones] that are supposed to be present in the hive. Certainly the brood smells like fresh bread, whereas the queen smells like lemon balm. There is also a so-called hive scent. But the viewpoints regarding the significance of these scents are in many respects inadequate. It is mistaken, for example, to consider that hive scent has the power to form a community; at most it is the reason why bees with a different scent are excluded. The queen's scent may indeed exercise an attracting stimulus; swarming bees will even surround a dead queen! Scattered swarms will also hover for days around the spot where the queen had landed.

But it is easy here to fail to recognize the deeper situation, for the significance of the queen's scent is not simply that the bees cluster around the queen because they are 'attracted by the scent', but that it creates a pathway to the bees for the group-soul, so that they become a colony. More important than the scent itself are the emotional responses stimulated by it in the bees. Where an absolutely real emotional stream of sympathy flows towards the queen, the queen-bound group-'I' has the possibility of taking hold also of the worker bees through these waves of feeling. As a rule this will always be the case with bees where the unity of the colony is loose, as for example in swarming bees and afterswarms. Fertilized queens receive the bees' sympathy more readily than unfertilized ones. Queenless colonies that are already producing drone brood—in other words, where the reproductive forces in the worker bees have already revived due to the absence of a queen—usually meet a fertilized queen with antipathy, stinging her to death. Unfertilized queens are also rejected by colonies that have just lost their fertile queen; and they will only accept a fertilized queen if she is introduced (with great caution) some hours after the previous queen has been removed so that a loosening of colony cohesion has already set in.

The great success of the old heather-region beekeeping was not only due to the intimate understanding its beekeepers had for colony life. They also knew—if only in an instinctive, feeling way—that colonies had to be treated as individuals, and they understood how this individual came into being. A subtle sense told them when an intrusive measure could be taken without causing damage to the beekeeping, because the periods of loosened colony cohesion were known to them.

We more intellectual, modern people need to develop a new instinct for these things. This is possible if we take thoughts to heart of the kind presented here.

The Question of Appropriate Beekeeping Techniques

Few discoveries have had such far-reaching impact on modern bee studies than the discovery that when a colony loses its queen—whether by dying from old age or through removal by the beekeeper—it produces new queens from eggs and young larvae that would otherwise have become workers, by converting their worker cells into queen cells, and feeding them protein-rich royal jelly.

Nickel Jakob already knew this fact. In his 1568 book on bees he writes that he treated newly queenless colonies by taking a piece of comb containing open brood from a strong colony, and incorporating it into the sick colony. In the 1614 Höfler edition of the same book he says: 'If the queen dies, one should help by adding brood so that they produce [a new] one.' Instructions are even given on how to carry out this manipulation.

This discovery gained wider practical significance in the age of frame hives, where the moveable comb frames allowed the entire mode of operation to be geared to this.

The old swarm method is widely unpopular today in beekeeping circles. The view is that the act of swarming is effectively only a renewal of the queen, and this can be achieved more comfortably and advantageously by other means, whether by making nucleus colonies or by artificial breeding of queens. The most extreme developments can be found in the swarmless methods of Russian apiaries, where brood is removed as soon as it is capped, kept in an incubation chamber, and used later in 'assembling' colonies. The operation is very simple: once the bees have emerged from them, a number of combs are put into a hive; a queen is introduced, and the 'colony' is ready. The aims even go as far as fertilizing the queen in laboratory conditions by introducing a shot of drone sperm into her by injection. Hitherto this has not had much success, but it shows where materialism in bee studies and bee breeding leads: the bee breeder becomes an engineer assembling honey-producing machines.

Anyone who sees a bee colony as just a mass of bees with a queen, will inevitably end up on the path of artificial methods because—as the Gerst school with its 'organic perception of the bee' says of artificial queen breeding—the course of events shows that 'it works'. Whether it will still work after 50 years is of no concern as long as there is a profit.

When a bee colony, for whatever reason, loses its queen and raises a new one from worker brood, this is initially an act of self-healing comparable to regenerating a lost limb. A relatively undifferentiated structure (an egg or young larva) is taken, and this, through the inner dynamic of the (group-soul oriented) totality, is specialized to the pattern of the missing part.

Anyone who thinks it possible to base and maintain beekeeping on the forces of self-healing per se, either has no sense at all for the totality, however constituted, of the hive, or when talking of 'the spirit of the bee' and 'the soul of the colony' is doing so merely in abstract or aesthetic terms. A group-soul as understood by spiritual scientific empiricism has conditions, as we have shown, under which the appropriate influence on the outer colony can take place. And only practices that take these conditions into account can be deemed appropriate. The ideal which the bee-keeper should strive for is not that of a cold-hearted bee engineer—though in its noblest form this may have a certain justification in the limited field of experimental science—but rather of the bee psychologist in the broadest sense, who, particularly out of a selfless warm-heartedness, is able to understand the group-soul nature of bee colonies even if only in a feeling and instinctive way to begin with. What has been presented in the previous section is intended entirely along these lines, and can be of service in this direction. For a foundation of good practice and the schooling of conscious judgement, the following can be said.

During the main spring inspection, beekeepers often finds marked differences in strength among the colonies. There are a number of practical measures that can give the weak colonies a boost. The simplest is relocation—a weak colony's site is exchanged with that of a strong colony. In this way it gets more forager bees than it gives out, since forager bees within their flight radius return to their original spot (provided they are not swarming).

Another measure, used in skep beekeeping, is a feed-and-move method. A plate smeared with honey is placed under a strong colony. As soon as the plate is covered with bees—it will mainly be young bees—the plate is placed under the weak colony. The young bees, now full of honey, are accepted without more ado into the foreign colony and remain there provided they have not already flown back to their original site.

A third method, possible with frame hives, is to take brood comb from a strong colony and put it into a weak one—but of course not more than the weak colony can manage.

In general there can be no objection to these practices which, mathematically speaking, amount to adding and subtracting. There is more advantage however in strengthening colonies using brood comb and by feeding and moving young bees rather than by relocation, because as long as young bees are not fully matured (the limit lies at roughly the time when they take up guard duty—at round 18 days according to Rösch's research), they are not yet 'fixed' in the hive, and can be accepted into another colony without friction. In the relocation method on the other hand, the strengthening is done through forager bees. These are more strongly aware of the change and initially more hesitant to join a different colony. But they also finally acclimatize to the new situation, and all the better and quicker when there is a strong 'call' from the forage when the sense of belonging to a colony is then 'drowned out' by foraging zeal. But one can also at times observe how bees of relocated colonies fly from flight hole to flight hole in the apiary until they find their colony again.

In summary we can say: Where the adding and subtracting methods in beekeeping are used in moderation, there may be occasional small incidents of friction among the bees; but the breeding itself is not undermined since the group-soul-bearing queen and the majority of the bees remain untouched. These methods can therefore be deemed as appropriate.

Methods whereby a colony is split into two or more parts, or a number of colonies are merged into one, can be described as methods of intensified addition and subtraction—or therefore of multiplication and division. It is more difficult in these cases to form a really professional judgement as to the appropriateness of these practices for beekeeping, for here it is through arbitrary human intervention that colonies are brought into existence or have it taken away; for it also happens that, particularly in large apiaries, concurrent swarms merge together. As mentioned above, colony cohesion is loosened in swarming bees; they are not 'fixed' to the full extent. The same applies in general to bees of colonies that are just about to swarm, and particularly to the bees left behind after a primary swarm. Similar swarms—that is two primary swarms or two afterswarms—usually merge without difficulty. Dissimilar swarms (primary and afterswarms), provided they have not already fought each other on merging, will only reach harmony and unity if the old queen is still alive. The bees of a primary swarm will hardly ever encounter an unfertilized queen with sympathy.

Deliberate merging of colonies by beekeepers is done mainly in spring and autumn—in other words, at times of minimal brood. Reduced egg laying by the queen is a sign that the reproductive drive in the colony is subdued. This means nothing other than that, because it is not so necessary, the suppression of the reproductive instinct in the worker bees is also weaker, which in turn amounts to a lower level of colony cohesion. Merging in early spring rather than in autumn has the advantage that during the period of increasing brood bees are assimilated more easily; the queen asserts herself and quickly establishes a colony body in which there are no longer any foreign components. Colonies that are merged in the autumn are less settled during the winter and have a stronger tendency for individual bees to fly off. Ramdohr's experiments at the beginning of the last [the nineteenth] century showed that trying to strengthen a colony in autumn is fairly futile. At best it is a measure of last resort to try to utilize a queenless or 'naked' colony after emptying the skeps in the heather regions. But this is not an entirely appropriate practice. Apart from individual instances that are unsuccessful, the possible disadvantages for the bee husbandry are negligible, particularly where in general appropriate methods are employed including, for example, the swarm method.

A colony should only be split when it is just about to swarm or, better still, when it has already cast its primary swarm. The remaining bees, which contain many young bees or developing queens, are good for creating nucleus colonies. By contrast, it should always be regarded as a brutal intervention of very dubious beekeeping merit when, in the middle of the period of increasing brood (around the first half of May), parts are extracted as nucleus colonies, forcing the raising of a new queen from young worker brood. A bee colony raising a queen for its own reasons, or being forced to by emergency circumstances, is definitely not the same thing. For the devotion to the new queen, and therefore also to the permeating power of the group-soul of the colony developing from the nucleus, is dependent on the manner in which the old conditions were dissolved. Concerning this, we can say the following.

We need to bear in mind that the group-'I's scope of influence is both psychological and physiological: the bees are not only 'emotionally' engaged, but the reproductive instinct in them is also suppressed. In the situation where a bee colony, out of itself, begins raising young queens—in other words, is preparing to swarm—the 'swarm doziness' that accompanies this is a sign that the emotional influence of the old group-'I' is being held back, while the physiological influence that keeps the reproductive instinct in check is barely affected, or not at all.

When, in a nucleus colony arbitrarily created by the beekeeper, the bees are physically removed from the old queen, the influence of the group-'I', hitherto engaging the feeling of the bees, suddenly ceases. And the cessation of the physiological influence is sensed even more strongly, for the physical changes triggered in these circumstances do not contribute to the bees' wellbeing in any way. In addition to this now-missing element which causes discontent in the bees, the chaoticizing radiations emitted by the developing queens is scarcely a relief and do not induce sympathy towards the queen, for it is predominantly the direct soul-influence of the old group-'I', assuming it is still present, that would be affected by this. But a reserved attitude of the bees towards the queen will necessarily lead to the new group-'I's grasp of the colony remaining weak. The practice of creating nucleus colonies therefore often results in colonies in which 'a certain spark' is missing.

To summarize; the following can be said of the multiplication and division methods consisting of all those using splitting and merging of colonies: they are appropriate and right when carried out at times of colony dissolution or of minimal brood, and when used to remedy an emergency situation (queenlessness, helping weak colonies). They were used in these ways and circumstances by the master beekeepers of the old skep-hives, who fared well with them.

In recent times other methods have been added to those mentioned above, which, keeping to the terminology we have been using, could be called the [mathematical] root and exponential methods.

Many beekeepers are dissatisfied with the performance of their colonies and blame the queen or her race. They therefore take to breeding queens 'of particularly good stock', or get 'pedigree' queens from breeder X from stock Y and certification Z. As soon as the hoped for 'prize' queens are acquired, the old ones are unceremoniously ejected, in order in this way to remove the problem 'at the root'. The colonies taken to their root in this way then get, amid all sorts of precautionary measures, the 'higher-performing' queens added to them to function as an exponent. We can ask, however, whether the powers of these 'exponents' are not more imaginary than real, and the advantages of the 'improved' colonies more in the imaginations of interested parties. The initial increase in activity seen in colonies that have had their queens exchanged for Italian queens, is not really so significant; for it is perfectly possible for the stimulus here to come from the new and unusual conditions in climate and forage. And the offspring are not renowned for their special qualities.

The arbitrary treatment of queens would not have increased to the extent that it has if, consciously or unconsciously, an atomistic view of the bee colony were not so prevalent in which the queen is purely an 'egg-laying machine'. The perception of the colony merely as an aggregate does not do it justice; for all the parts making up the sum derive the possibility of their existence from the colony as a totality, and each one taken by itself is not a reality. If we wish to grasp the hive mathematically as an organism permeated by the power of a group-soul, then all the formulae which might be applicable for an external perception of the colony require a supplementary element. In particular, in a mathematical grasp of the beekeeping measures mentioned above, the loosing and linking of the soul-spiritual bond needs to be taken into account. When this is done, one comes into the realm of differential and integral calculations.

The equation for the performance of a colony viewed atomistically, could be expressed by the formula:

$$y = f\,(a + b + c)$$

This states that the colony's work potential (y) is equal to the function (f) from the queen's own forces (a), the worker bees (b) and the drones (c).

The supplement of this equation, in order to encompass the colony as a whole, must be one in which, among other things, the integral symbol ($\int$) is placed at the beginning and the symbol for the impulse deriving from the group soul (dx) is added.

$$y = \int f(u, v, w)\, dx = F(x)$$

In this formula, u, v, w are to be regarded as complicated, inter-related functions relating to the queen (u), worker bees (v) and drones (w). The symbol F (x) indicates that the outer colony is an organ of the group-soul (x).

Since in the genesis of a colony the queen is the primary bearer of the group-soul, with the workers having a secondary connection to it through their devotion to the queen, it becomes obvious that the inner cohesion of a colony in general will be the stronger the more closely related the bees are to the queen. For bee husbandry that is not sooner or later to be ruined, this fact is essential; it decides the question of race in favour of indigenous bees.

Regarding operational practices, it follows that any beekeeping measures affecting the queen also affect the colony as an individual in the deepest and most drastic way, and therefore should be kept to an absolute minimum. Any arbitrary changing of queens should be categorically avoided. Where it is necessary to give a new queen to a colony that has lost the old one, or to queenless bees remaining behind after a swarm, then rather than introducing a queen alone, add a small nucleus colony taken from left-behind bees or a small secondary swarm. Anything beyond this, i.e. the whole practice of artificially breeding queens, together with the swarmless methods, is a breeding nonsense that will one day take its revenge.

Any violation of the conditions under which the group-soul can exert its influence to the full, undermines the communal life of the colony. And if this should diminish in the course of time even only to the point where, following a weakening of group-soul influence, the bees, as with the Egyptian bee, did not form a winter cluster and the reproductive instinct in the worker bees was so minimally suppressed that it immediately revived when the queen was lost, then this would be enough to render beekeeping in our latitudes impossible. The colonies would die of cold in the winter like the bumblebees. This last point would, however, make no difference to the fate of beekeeping since no yield of any note could be attained on this basis. Even today it is the case that good yields under normal foraging conditions are only to be got from colonies that are both inwardly and outwardly strong.

In no way does one have to be a reactionary to reject artificial queen breeding and favour swarm methods, for it is possible to practice queen breeding in moderation and in an appropriate way, as follows:

The primary swarm, in which the old colony (as per group-soul) continues, should be accepted and not, as has become fashionable in many quarters, put back after removing the queen. For the bees that left with the old queen will not easily make a connection to the young unfertilized queen, and they are not hard workers. As a new colony in itself however, primary swarms are extremely industrious. If they are too small, they are easily improved by adding brood.

In most cases, if remaining bees are left to develop undisturbed, two or three secondary swarms will be cast. To establish these as separate colonies would only be useful if they are early and the main forage is heather. It is not always necessary to leave things to nature in order to observe appropriate and right methods, for it is in accord with a basic and healthy principle to stop splitting by creating one new colony from left-behind bees. There are two ways to do this: either let the secondary swarm occur, but put it back in the evening after having removed all the queen cells in the hive; or, on the seventh or eighth day after the primary swarm (the new queens are usually piping and croaking by now) remove all the queen cells except one mature one. If with this manipulation some queens have already emerged—which is often the case—then all the queen cells can be removed and one or two of the emerged queens left. The bees will not now swarm and will themselves remove the superfluous queens, of which there are often several also in a secondary swarm. If the bees are left to choose for themselves which queen to keep, it is only in the rarest cases that all the queens will be killed. Often one finds superfluous queens in a cluster of bees on the floor of the skep or in a corner of the box. It is advisable to remove these but only when there are no symptoms of queenlessness (agitated running about and searching in front of the flight hole). Depending on circumstances, both methods have their advantages and disadvantages.

In a large apiary it is best to use remaining bees as nucleus colonies, creating reserve colonies to re-queen queenless ones or late left-behind bees.

This mode of operation is not only appropriate but also economic. It enables many colonies to be looked after by one person. And when things are managed so that, once swarming is over and before the main foraging begins, the majority of colonies have their rightful queen—a situation always to be striven for—then the honey yield can be good, if nature does not fail with her blessings of nectar.

Afterword

The ancestors walked as they sang, and when it was time to stop, they slept. In their dreams they conceived the events of the following day, points of creation that fused one into another until every creature, every stream and stone, all space and time became part of the whole, the divine manifestation of the one great seminal impulse.

Wade Davis, *The Wayfinders*

In his multifold exposé of *The Spiritual Foundations of Beekeeping*, Lorenzen embarks on a voyage into the world of honeybees, sowing seeds for a new approach to apiculture. With every page we turn, this voyage metamorphoses into a pilgrimage to the sacred life-scape of honeybees. A new fabric of a paradigm emerges as he reframes the fundamentals of our relationship with the phenomenon we call honeybees and their significance within the world.

Lorenzen reveals the ancient cultural and spiritual wisdom of the seemingly lost 'time-being' of our ancestors and lays the foundation of a spiritual renewal in the context of apiculture. His approach recalibrates all conventional references, the field of honeybees, and even our own identity. Linked to the ground of being, it is a homecoming to our 'original face' before birth and a recognition of the spiritual dimension of reality. Lorenzen lays the groundwork for the liberation of honeybees from what is called 'beekeeping' and introduces a new framework of interrelationship and dignity.

When we follow Lorenzen's overview into the genesis of the honeybees, it may feel as if we were entering a magical, foreign, or forgotten world. It is a world of spiritual clarity and wisdom-filled intimacy with the life forces of the universe. In this context, we learn about the mystery schools and communication skills to work on an archetypical level, tapping into a fluidity of being and of transforming a fig wasp into the image of a honeybee. Epigenetics just scratches the surface of the image of such plasticity, and quantum biology may feel a deep resonance.

As the default image of honeybees transforms throughout this exploration, our altered perception forms new images within language as well. These images can precipitate into a new terminology, with the intent to reflect another reality of being. The linguistic reflection of the phenomenon of honeybees could also be expressed as something I refer to as *onebeeing*. It is a play with words and an internal exercise of openness and receptivity. Unlike our own default sense of self, the sense of self of the *onebeeing* is not defined as a living being separate from the world, but rather as being part of the world through intimate belonging. Her sense of self is created in dynamic and multidimensional processes, as if she were embodying a network of multiple selves. We could almost say that she 'are' more other than self. This cornucopia of levels of self also wants to be reflected linguistically and therefore can create a renewed intimacy with our own language and mother tongue. The many bees appearing to our eyes 'is' a time-gestalt, a gesture of the Great Bee. She 'are' a trans-vertebrate, positioning herself outside the typical biotic classifications, and as such bewilders our current cosmology of rational thinking. She comes with a fluidity and plasticity on all levels of being, and the closer we look, the less we can place her into categories. As long as we are limited by a dualistic analytical thinking, her true nature will stay elusive.

Among the fundamental characteristics of the *onebeeing*, inwardness is, according to Lorenzen, one of her most essential attributes. Imbued with soulfulness and wisdom, this inwardness is an interface for communication and a medium to reconnect with the sacred in all life forms. It holds the key to a comprehensive understanding of the *onebeeing* and is a seed for awareness and contemplation. What can follow is a personal and ingenious impulse to create a ceremonial field, in which apiculture becomes a transformative practice, unveiling the phenomenon of the *onebeeing* as a spiritual gate.

Lorenzen portrays the emanation of the *onebeeing* as a cosmic being, and shows how she conveys an enlightened gesture of living in the world by exemplifying a sacred life of grand-motherly love and selfless giving for the benefit of all. An alchemical process becomes apparent, as life forces are being transformed to what he calls life-spirit. This life-spirit illuminates a warm and soulful unfolding of being in the world, and it epitomizes a radically different paradigm of life. It speaks of a vision of a new archetype of being in the world, as it relates to our current existential global and personal challenges. The *onebeeing* is soul plasma.

She is an embodiment of hope by example. Once we immerse ourselves in her life-scape, it will be self evident that contemplation and inwardness are not only essential when practising apiculture, but are keystone elements for a transformation of our life and a sustainable future on earth. As a sister being, the *onebeeing* 'are' a portal into our own future as a fellow social being. She is a gift given to us and reveals the deep truth of our life, soul, and spirit.

Michael Joshin Thiele

Notes

1. Goethe, *Goethes Werke*, Weimar: Hermann Böhlau, 1887–1919, II. Abtheilung: Naturwissenschaftlichte Schriften, Bd. 4, pp. 295–296, translation according to Wikipedia, accessed 5/12/16.
2. Goethe, *Scientific Studies*, p. 307 as referenced in *Goethe's Way of Science: A Phenomenology of Nature*, Seamon and Zajonc ed., State University Press of New York, 1998.
3. Goethe, *Goethe's Colour Theory*, R. Matthaei ed., van Nostrand Reinhold, New York, 1971, p. 57.
4. *The Wholeness of Nature: Goethe's Way of Science*, Bortoft, Floris Books, 1996, p. 22.
5. Ibid.
6. Charles Martin Simon, *Principles of Beekeeping Backwards*, Bee Culture, July, 2001.
7. Steiner, *Bees, Lecture by Rudolf Steiner*, Anthroposophic Press, 1998, p. 69.
8. W. Mao, M. A. Schuler, M. R. Berenbaum, *A dietary phytochemical alters caste-associated gene expression in honey bees.* Sci. Adv. 1, e1500795 (2015).
9. Prologue to *The Buzz About Bees, Biology of a Superorganism*, Jürgen Tautz, Springer, 2008.
10. The connection of morality and body odour is an ancient one. Aristotle considered it, and Christian culture also held that holiness was linked with bodily fragrance. 'In Christian culture, corporeal fragrance was not, or should not be, an indication of good grooming but of good morality.' – see: http://www.encylopedia.com/doc/1O128-bodyodours.html

    This was apparently also the case in Sufi tradition where '... the link between holiness and the body likewise grants the bodies of saints a sweet-smelling scent, whereas the bodies of evil or avaricious individuals exude an unpleasant odour. Again, this belief is paticularly prevalent among the Sufis, who speak of sweet smells (rather than the putrefaction of decomposition) emanating from the tombs of saints. Living saints as well are said to have blood that is pure and not to exude offensive body odour, in contrast with wordly people, whose blood is foul-smelling.'—see: *Embodiment, Morality, and Medicine* edited by L. S. Cahill, M. A. Farley.

A recent study has shown that certain character traits can be assessed by smell alone. See:
http://onlinelibrary.wiley.com/doi/10.1002/per.848/abstract

However, the background to Lorenzen's connecting this with the bees and flowers is obscure.

11. Not only pollen but the whole flower can maintain a temperature higher than the surroundings. 'The flowers of some plants produce enough heat to raise their temperatures as much as 35°C above air temperature. Three species have been shown to regulate flower temperature within a narrow range by an unknown physiological mechanism that increases the rate of heat production as air temperature decreases.' (*Heat-producing flowers* by Roger S. Seymour and Paul Schultze-Motel.) See
http://instructional1.calstatela.edu/kfisher2/BIOL360/ecology.pdfs/hot.flowers.pdf
12. Calcium carbonate (found in chalk, limestone and marble) has a well-known neutralizing effect on acids. Examples are its use as an antacid in indigestion pills; its application as lime to reduce soil acidity; and many industrial uses. In crystal form—calcite—it has a light-splitting property or double refraction, known as birefringence. Lorenzen is noting that these calcium carbonate particles in the intestinal wall, are at a point where acid nutrients must be neutralized when passing ino the alkaline blood. The fact that they are absent in bees affected with nosema is significant.
13. The excellent dissertation of Ravasini on *The Fig Trees of Italy* (Bern, 1911) was used for the following presentation. The illustration of figs and fig wasps are also from that work.
14. H. Müller, *Anwendung der Darwin'shen Lehren auf Bienen* [Applying Darwinian Studies to the Bee]. Verh. d. naturh. Vereins d. pr. Rheinlande, 1872.
15. H. v. Buttel-Reepen, *Die stammesgeschichtliche Entstehung des Bienenstaates* [The Origin and Evolution of the Bee Colony]. Leipzig 1903.

16. Detailed description in H. Friese, *Die europäischen* [The European Bees]. Berlin and Leipzig, 1923.
17. From H. v. Buttel-Reepen: *Leben und Wesen der Bienen* [The Life and Nature of Bees]. Braunschweig 1918.
18. From Friese ibid.
19. See Friese ibid.
20. *Die Temperaturregelung im Bienenvolk.* [Regulation of Temperature in the Bee Colony.] Zeitschr. f. vergl. Physiologie 1926.
21. Lectures on 26 Sept. 1905, 29 Sept. 1905, 14 Sept. 1907, 2 Feb. 1908; see also *From the Akaschic Chronicle.*

A note from the publisher